Checklisten der Fauna Österreichs, No. 8

Hieronymus DASTYCH:

Tardigrada

Werner E. HOLZINGER, Andreas CHOVANEC
& Johann A. WARINGER:

Odonata (Insecta)

Herausgegeben von Reinhart Schuster

Serienherausgeber
Hans Winkler & Tod Stuessy

**VERLAG DER
ÖSTERREICHISCHEN
AKADEMIE DER
WISSENSCHAFTEN**

Titelbild: *Hypsibius klebelsbergi* (MIHELČIČ, 1959). — Eine ausschließlich auf Gletscher vorkommende (= kryobiontische) Bärtierchen-Art, die bisher nur aus Österreich bekannt ist (Pseudoendemit?); Langtaler Ferner, Ötztaler Alpen; (Foto: R. WALTER & H. DASTYCH, Zoologisches Museum Hamburg).

Layout & technische Bearbeitung: Karin WINDSTEIG

Checklists of the Austrian Fauna, No. 8. Hieronymus DASTYCH: Tardigrada. Werner E. HOLZINGER, Andreas CHOVANEC & Johann A. WARINGER: Odonata (Insecta).

ISBN 978-3-7001-7600-8, Biosystematics and Ecology Series No. 31, Austrian Academy of Sciences Press; volume editor: Reinhart SCHUSTER, Institute of Zoology, Karl-Franzens-University, Universitätsplatz 2, A-8010 Graz, Austria; series editors: Hans WINKLER, Austrian Academy of Sciences, Dr. Ignaz Seipel-Platz 2, A-1010 Vienna, Austria & Tod STUESSY, Herbarium, Museum of Biological Diversity, The Ohio State University, 1315 Kinnear Road, Columbus, Ohio 43212, U.S.A.

A publication of the Commission for Interdisciplinary Ecological Studies (KIÖS)

Checklisten der Fauna Österreichs, No. 8. Hieronymus DASTYCH: Tardigrada. Werner E. HOLZINGER, Andreas CHOVANEC & Johann A. WARINGER: Odonata (Insecta).

ISBN 978-3-7001-7600-8, Biosystematics and Ecology Series No. 31, Verlag der Österreichischen Akademie der Wissenschaften; Bandherausgeber: Reinhart SCHUSTER, Institut für Zoologie, Karl-Franzens-Universität, Universitätsplatz 2, A-8010 Graz, Österreich; Serienherausgeber: Hans WINKLER, Österreichische Akademie der Wissenschaften, Dr. Ignaz Seipel-Platz 2, A-1010 Wien, Österreich & Tod STUESSY, Herbarium, Museum of Biological Diversity, The Ohio State University, 1315 Kinnear Road, Columbus, Ohio 43212, U.S.A.

Eine Publikation der Kommission für Interdisziplinäre Ökologische Studien (KIÖS)

Inhalt

Hieronymus DASTYCH

Tardigrada

Summary 1
Zusammenfassung 1
I   Einleitung 2
II Allgemeiner Teil 3
    1. Erforschungsgeschichte und aktueller Forschungsstand in Österreich 3
    2. Datengrundlage 3
3. Aktueller Forschungsstand 4
III Spezieller Teil 5
    1. In die Liste aufgenommene Arten und Unterarten 5
    2. Problematica 19
IV Literatur 20

Werner E. HOLZINGER, Andreas CHOVANEC & Johann A. WARINGER

Odonata (Insecta)

Summary 27
Zusammenfassung 27
I   Einleitung 28
II Allgemeiner Teil 28
    1. Erforschungsgeschichte und aktueller Kenntnisstand 28
    2. Biologie der Libellen 30
    3. Libellen als Bioindikatoren 34
III Spezieller Teil 36
    1. Systematik und Nomenklatur 36
    2. Angaben zu den Arten 36
    3. Kommentiertes Verzeichnis der Libellen Österreichs 37
    4. Irrtümliche und fragliche Meldungen 47
    5. Exotische Arten in Gewächshäusern 47
IV Literatur 48

Vorwort

In der vorliegenden Folge werden zwei Gruppen wirbelloser Tiere, wie sie unterschiedlicher nicht sein könnten, bearbeitet – die Bärtierchen (Tardigrada) und die Libellen (Odonata).

Obgleich die Tardigrada eine so geringe Körpergröße wie manche Einzeller haben, ist ihr kleiner Körper nicht nur aus einer Zelle, sondern aus verschiedenen Zellen und sogar Organen aufgebaut. Ihre 8-beinige Gestalt ist eigenartig, wie das elektronenmikroskopische Titelfoto instruktiv zeigt. Auch die verborgene Lebensweise (vorwiegend in Böden) trägt dazu bei, dass die mikroskopisch kleinen Tardigraden nur wenigen Leuten bekannt sind. Es wird daher mit Überraschung registriert werden, dass immerhin fast 80 verschiedene Arten bei uns vorkommen.

Im Gegensatz dazu stehen die im zweiten Beitrag bearbeiteten Libellen. Es sind insgesamt 78 Arten, die in Österreich anzutreffen sind. Sie gehören mit zu den größten und buntesten Fluginsekten unserer Heimat und sie zählen daher zu den allgemein bekanntesten heimischen Wirbellosen. Dies prägt sich auch dahingehend aus, dass in der Bevölkerung auch deutsche Gruppen- und Artnamen im Gebrauch sind. Aus diesem Grund sind im vorliegenden Checklisten-Beitrag ausnahmsweise auch deutsche Bezeichnungen zusätzlich genannt.

Reinhart SCHUSTER
Bandherausgeber

Tardigrada

Hieronymus Dastych

Summary: Seventy-nine species of Tardigrada, including 14 new reports, have been recorded within Austria. Of the 133 taxa previously reported by various authors, 41 species represent nomina dubia, nine are synonyms that have been previously recognized, eight species still need to be revised further or their reported occurrence within the country needs to be confirmed, three species are nomina nuda, eight species represent species groups containing cryptic taxa, and two species have been misidentified. The majority of data on tardigrades from Austria has been provided by F. Mihelcic between 1949 and 1972, but these data are largely confusing, and hence, of limited utility. Most of the species (63) have been reported from the province of Tyrol. The tardigrades of Austria are still poorly known, and a modern revision is urgently needed.

Zusammenfassung: Aus Österreich sind 79 valide Tardigradenarten bekannt; davon sind 14 Arten neu für die heimische Fauna. Es sind insgesamt 133 Taxa gemeldet, davon sind 41 Arten (!) nomina dubia, neun sind bereits früher synonymisiert worden; drei Arten sind nomina nuda, acht sind revisions- resp. bestätigungsbedürftig, weitere acht sind kryptisch und zwei sind eindeutig falsch bestimmt. Die meisten Daten im Zeitraum 1949–1972 wurden von F. Mihelčič geliefert, doch sind diese, wenn überhaupt, nur bedingt brauchbar! Die meisten Arten (63) sind im Bundesland Tirol nachgewiesen. Insgesamt ist die Datenlage uneinheitlich. Die Tardigradenfauna Österreichs bedarf dringend einer sorgfältigen Revision und Neubearbeitung.

Key words: Tardigrada, Austria, Checklist, Biodiversity

I Einleitung

Tardigrada („Bärtierchen") sind winzige, vorwiegend zwischen 0,2 und 0,5 mm große vielzellige Tiere (Metazoa) mit walzenförmigem Körper, undeutlich abgesetztem Kopf, vier Rumpfsegmenten und acht, mehrheitlich stummelförmigen Laufbeinen, die bei vielen Arten mit Doppelkrallen enden (z. B. GREVEN 2013). Tardigraden sind limno-terrestrisch, d.h. sie leben in marinen und limnischen Habitaten, bei ausreichender Feuchtigkeit jedoch viele Arten auch in gut durchlüfteten terrestrischen Habitaten wie Moosen, Lebermoosen, Flechten, verrottendem Laub und im Boden selbst. Dort ernähren sie sich von Bakterien, Algen, Pilzen, dem Inhalt von Mooszellen, Detritus sowie einzelligen Tierchen (Protozoen), Fadenwürmern (Nematoden), Rädertierchen (Rotatorien) und auch Artgenossen. Zurzeit sind über 1150 Arten beschrieben (u. a. GUIDETTI & BERTOLANI 2011). Diese werden in drei Klassen gegliedert, und zwar in die Heterotardigrada, Mesotardigrada und die Eutardigrada. Tardigraden sind weltweit verbreitet; sie kommen von der Tiefsee (Abyssal) bis in schneebedeckten (nivalen) Zonen der höchsten Bergen vor.

Viele „terrestrische" Tardigraden überdauern die Austrocknung ihres Lebensraums in einen kryptobiotischen Zustand (= Zustand latenten Lebens), der als Anhydrobiose bezeichnet wird. Sie bilden Dauerstadien („Tönnchen"), die außerordentlich resistent selbst gegen extreme, nicht auf der Erde vorkommende Umwelteinflüsse sind (z. B. BAUMANN 1922, WRIGHT 2001). Seit ihrer Entdeckung wird über die Stammesgeschichte dieser offenbar sekundär vereinfachten Organismen gestritten. Allerdings kamen schon zahlreiche frühere Autoren nach sorgfältiger Prüfung aller bis dahin bekannten Merkmale der Tardigraden zu der Auffassung, Tardigraden seien verwandt mit den Gliederfüßern (Arthropoda) (u. a. PLATE 1888, BASSE 1905, MARCUS 1929). Interessanterweise stützen auch neuere und neueste morphologische und molekularbiologische Daten diese Meinung, oft ohne die Argumente und die Meinungen der früheren Autoren gebührend zu berücksichtigen. Über die Stellung der Tardigraden innerhalb der „Arthropoda" wird aber immer noch gerätselt (Zusammenfassung bei MAYER et al. 2013).

Danksagung: Ich danke ganz herzlich allen Personen, meistens aus dem Deutschen und Österreichischen Alpenverein, die freundlicherweise für mich Proben aus Österreich zur Verfügung stellten: Hanna DORENBURG, Prof. Dr. Ernst EBERMANN, Thorsten GROTHKOPP, Horst HASS, Dr. Barbara KNOFLACH-THALER, Rose-Marie MAYR, MSc Oliver NÄGELE, MSc Barbara POST, MSc Lukas RINNHOFFER, Prof. Dr. Birgit SCHLICK-STEINER, Heinrich SIMON, †Prof. Dr. Konrad THALER, Dr. Florian STEINER und Matthias UPHUES. Weiteres danke ich Prof. Dr. Hartmut GREVEN (Universität Düsseldorf) und Prof. Dr. Reinhart SCHUSTER (Universität Graz) für ihre kritischen Anmerkungen und Verbesserungsvorschläge. Ich bin dankbar der Universität Hamburg und der Alpine Forschungsstelle Obergurgl (Universität Innsbruck: Prof. Dr. Brigitta ERSCHBAMER und Dr. Nikolaus SCHALLHART) für ihre Unterstützung.

II Allgemeiner Teil

1. Erforschungsgeschichte und aktueller Forschungsstand in Österreich

Erste Angaben über Tardigraden aus Österreich stammen von EHRENBERG (1853), der eine neue Art, *Echiniscus altissimus*, vom Massiv „Montis Gross-Glockner" beschrieb. Die ersten Bärtierchen in Bodenproben (Kufstein, Pyramidenspitze, Brentenjoch) wurden von FRANCÉ (1921) gemeldet. MARCUS (1928, 1930) beschrieb zwei neue Arten und listet auch einige andere Taxa auf (l. c. 1936). MICOLETZKY (1910/11), VORNATSCHER (1938: det. E. MARCUS) und KÜHN (1940) berichteten über einige limnische Tardigraden, die sie bei Salzburg und Wien gesammelt hatten. FRANZ & MIHELČIČ (1954) meldeten 42 Arten (det. F. MIHELČIČ) aus den Nordost-Alpen, STEINBÖCK (1957) berichtete über das Vorkommen von Tardigraden in Kryokonitlöchern an der Gletscheroberfläche der Stubaier und Ötztaler Alpen.

Die heutigen Kenntnisse über österreichische Bärtierchen – überwiegend aus Osttirol, Kärnten, Steiermark und Salzburg – sind vor allem F. MIHELČIČ (1949–1972) zu verdanken. Der Autor beschrieb aus Österreich 37 neue Taxa und meldete von dort zahlreiche bereits bekannte Tardigraden (die Auflistung aller Veröffentlichungen von MIHELČIČ siehe KOFLER 1978). Der Wert dieser taxonomischen und faunistischen Informationen leidet unglücklicherweise an der meist ungenauen und ungenügend dokumentierten Arbeitsweise des Autors. Dies betrifft besonders die Beschreibung von neuen Arten. Dies muß natürlich Auswirkungen auf das derzeitige Bild der österreichische Tardigradenfauna haben, das zwangläufig noch sehr lückenhaft und unvollkommen ist (siehe Problematica).

Weitere faunistischen und taxonomische Informationen über Bärtierchen aus Österreich haben im Laufe der Zeit JANETSCHEK (1957), IHAROS (1966), TILZER (1968), MAUCCI (1974), FRANZ (1950: 25 Arten: det. F. MIHELČIČ; 1975: 13 Arten: det. G. IHAROS), (1983), KRAUS (1977), SCHUETZ (1987), THALER (1999), HOSCHITS (2004), DASTYCH & THALER (2002), DASTYCH et al. (2003), KHIEL et al. (2007), DASTYCH (2005, 2007, 2009a, b, 2011), POST (2012) und DABERT et al. (2014) zusammengetragen.

2. Datengrundlage

Als Grundlage für die vorliegende Liste dienten alle mir zugänglichen Literaturangaben mit Datensätzen über gemeldete Arten und deren Fundorte innerhalb Österreichs. Diese Literatur ist relativ weit verstreut und sehr unübersichtlich, da sie zahlreiche (oft nicht als solche gekennzeichnete) Wiederholungen (MIHELČIČ, FRANZ) enthält (s. u.). Darüber hinaus umfasst die Literaturliste aber auch Publikationen, die nicht im laufenden Text zitiert werden, aber deren Inhalt für Angaben über das Vorkommen von zitierten Arten in entsprechenden Bundesländer benutzt wurde. Die Information über neue Arten für die österreichische Fauna und manche schon aus dem Bundesgebiet bekannte Arten beziehen sich auf noch nicht publiziertes Material der Sammlung des Verfassers, die im Zoologischen Museum der Universität Hamburg aufbewahrt ist. Die Sammlung mit mikroskopischen Präparaten von F. MIHELČIČ (untergebracht im Tiroler Landesmuseum, Innsbruck), das einzige Belegmaterial für seine zahlreichen Meldungen über österreichischen Tardigraden, ist klein (35 Präparate), sehr schlecht erhalten und daher unbrauchbar! Von insgesamt 37 von MIHELČIČ aus Östereich neu beschriebenen als auch anderen von dort gemeldeten Tardigradenarten, ist nur von einer einzigen Art, *Hypsibius klebelsbergi*, ein Präparat (mit vier Syntypen) vorhanden, sodass wesentliche morphologische Merkmale nachprüfbar sind (siehe DASTYCH 1993).

3. Aktueller Forschungsstand

Aus dem heutigen Bundesgebiet wurden bisher 133 Tardigraden-Arten (bzw. Unterarten) gemeldet. Tatsächlich ist die Zahl von validen Arten aber deutlich kleiner, wenn man die Vielzahl von fragwürdige und problematische Taxa und Namen ausschließt. Zu diesen gehören zahlreiche nomina dubia (41 Arten!), mittlerweile synonymisierte Arten (9), revisions- oder/und bestätigungsbedürftige Arten (8), nomina nuda (3), Fehlbestimmungen (2 Arten) und kryptische Arten (8). Nach Abzug diesen Taxa (mit Ausnahme der krypischen Arten) bleiben 66 valide Arten, d.h. nur etwa 50 % von allen bisher aus Österreich gemeldeten Arten. Zählt man noch 14 für die österreichische Fauna neue Arten, hauptsächlich aus Nordtirol hinzu (unpubl., det. H. DASTYCH), kommt man auf insgesamt 79 Tardigraden-Arten, deren Vorkommen in Österreich nachvollziehbar ist.

Im „alten" Catalogus Faunae Austriae hat MIHELČIČ (1962a) 63 Taxa aufgelistet, wobei er allerdings drei von ihm früher gemeldeten Arten (*I. tuberculoides, D. gerdae, D. chilenense*) übersehen hat! Sein Katalog repräsentierte damals die erste gesamte Übersicht der Tardigraden Österreichs. Legt man jedoch die aktuelle, vorliegende Liste mit ihren kritischen Anmerkungen zugrunde, blieben von den 63 Arten nur 33 Arten übrig, die als valide angesehen werden können. Dies würde 42 % der jetzigen Artenzahl ausmachen und eine Verzweifachung der Artenzahl in den letzten 50 Jahren bedeuten.

Aus Österreich wurden bisher 47 neue (Unter-)Arten beschrieben (37 von MIHELČIČ selbst). Davon waren bereits vier als Synonyme bekannt (*Echiniscus clavisetosus, E. rosaliae, Cornechiniscus intermedius, Hypsibius callimerus*). Diese Arten sind überwiegend nur aus Österreich bekannt und oft nur von einzelnen Funden (siehe u. a. die Zusammenfassung von RAMAZZOTTI & MAUCCI 1983, MCINNES 1994, KOFLER 1978). Im am besten untersuchten Bundesland Tirol wurde die größte Zahl (63) von validen (= guten) Arten nachgewiesen (Nordtirol 35, Osttirol 41), eine etwas geringere Zahl in Kärnten (31), im Land Salzburg (27), in der Steiermark (21), in Niederösterreich und Oberösterreich (in beiden Bundesländern 20), im Burgenland (12) und die kleinste Zahl im Vorarlberg und Wien (in beiden Bundesländern nur zwei Arten). Diese Zahlen spiegeln in erster Linie die unterschiedliche Intensität wider, mit der Tardigraden in den verschiedenen Bundesländern gesammelt worden sind und nicht ihre tatsächliche Vielfalt.

In Österreich sind Bärtierchen mit den Heterotardigrada (nur Echiniscoididae) und Eutardigrada (Parachela und Apochela) vertreten, und zwar mit 10 Familien, 24 Gattungen und zwei Untergattungen. Die marinen Arthrotardigrada (innerhalb der Heterotardigraden) gibt es verständlicherweise nicht im Bundesgebiet.

III Spezieller Teil

1. In die Liste aufgenommene Arten und Unterarten

Die vorliegende Liste umfasst alle Taxa, die in verfügbaren Publikationen zitiert sind oder dem Verfasser als eigenes, noch nicht veröffentlichtes Material vorliegen. Die mehrmals publizierten (!) Neubeschreibungen derselben Art (= manche neue Taxa von MIHELČIČ) sind chronologisch aufgelistet. Die Synonymie-Quellen und der locus typicus werden nur für die Taxa angegeben, die als neue Arten aus Österreich beschrieben wurden. Die nicht mehr nachprüfbare Zugehörigkeit von einigen *Diphascon*-Arten zu den Untergattungen *Diphascon* oder *Adriopion,* die MIHELČIČ als neue Taxa vor der Revision von PILATO (1987) beschrieben hat, wurde subjektiv in der Liste als „(D.?)" oder „(A.?)" entsprechend markiert. Die supraspezifische Nomenklatur richtet sich vorwiegend nach den Zusammenfassungen durch PILATO & BINDA 2010, GUIDETTI & BERTOLANI 2011 und MARLEY et al. 2013.

Kürzel der Bundesländer (bzw. Landschaftsteile):

B   =   Burgenland
K   =   Kärnten
N   =   Niederösterreich
O   =   Oberösterreich
S   =   Salzburg
St =   Steiermark
T   =   Tirol (nT = Nordtirol, oT = Osttirol)
V   =   Vorarlberg
W   =   Wien

Weitere Abkürzungen und Bemerkungen:

l. c.	=	loco citato, an zitierte Stelle
l. cl.	=	locus classicus, ein alter Begriff für locus typicus
loc. typ.	=	locus typicus (Originalfundort)
*	=	Erstnachweis für betreffendes Bundesland (unpubl., det. H. DASTYCH)
Syn.	=	Synonym
Verbr.	=	derzeit bekannte Fundmeldungen in einzelnen Bundesländern
?	=	gemeldete, aber revisions- oder bestätigungsbedürftige Art
Fettdruck	=	vertrauenswürdig nachgewiesene Art, oder Arten, deren Vorkommen im Bundesgebiet sehr wahrscheinlich ist (siehe Problematica)

Klasse HETEROTARDIGRADA Marcus, 1927

Ordnung Echiniscoididae Kristensen & Hallas, 1980

Familie Echiniscidae Marcus, 1927

Parechiniscus Cuénot, 1926

Parechiniscus chitonides Cuénot, 1926
  Verbr.: nT*
  Erstnachweis für Österreich.

Bryodelphax Thulin, 1928

Bryodelphax parvulus Thulin, 1928
  Verbr.: K, N, O, St, oT
  (siehe Problematica)

Echiniscus Schultze, 1840

Echiniscus altissimus Ehrenberg, 1853
  Locus typicus: „Montis Gross-Glockner": Ehrenberg 1853; oT.
  Verbr.: oT
  Nomen dubium

?*Echiniscus arctomys* Ehrenberg, 1853
  Verbr.: oT

Echiniscus bigranulatus Richters, 1907
  Fehlbestimmung: Maucci 1974.
  Verbr.: oT

Echiniscus blumi RICHTERS, 1903
>Verbr.: K, S, St, oT

Echiniscus canadensis MURRAY, 1910
>Verbr.: oT

Echiniscus capillatus RAMAZZOTTI, 1956
>Verbr.: nT*
>Erstnachweis für Österreich.

Echiniscus granulatus DOYÈRE, 1840
>Syn. *Echiniscus (E.) abanti* MAUCCI, 1973: MAUCCI 1974.
>Verbr.: K, N, O, S, St, nT*, oT

Echiniscus jagodici MIHELČIČ, 1951
>Locus typicus: „der Höhe von 2100 m in der Laserz in Osttirol": MIHELČIČ 1951.
>Nachträglich (und widersprüchlich) erwähnt MIHELČIČ (1962a) „Kalsertörl l. cl." als loc.
>typ. für dieser Art.
>Verbr.: oT
>Nomen dubium

Echiniscus lapponicus THULIN, 1911
>Verbr.: nT*
>Erstnachweis für Österreich.

Echiniscus merokensis RICHTERS, 1904
>Verbr.: K, N, S, St, nT*, oT

Echiniscus nobilis MIHELČIČ, 1967
>Locus typicus: Werndorf: MIHELČIČ 1967a (St); andere Fundorte (l. c.): Doberdob
>(Italien: Triest), Delnice (Kroatien).
>Verbr.: K, St
>Nomen dubium

Echiniscus postojnensis MIHELČIČ, 1967
>(= „*E. postumiensis*" (sic!), „Adelsberg (Krain)": MIHELČIČ 1967b: 232; = lapsus linguae
>& nomen nudum).
>Locus typicus wurde in der Originalbeschreibung (l. c. 1967a) nicht präzisiert.
>Ursprünglich hat MIHELČIČ (1939) die Tiere aus der Umgebung von Adelsberg
>(= Postojna) beschrieben und als *Echiniscus loxophthalmus* Richters, 1911 (fehl)be-
>stimmt. Anhand zusätzlichen Materials aus Kärnten und Steiermark, hat er (1967a) das
>Taxon als eine neue Art (*E. postojnensis*) aufgestellt.
>Verbr.: K, St (Wildon)
>Nomen dubium

Echiniscus quadrispinosus RICHTERS, 1902
>Syn. *Echiniscus scrofa* RICHTERS, 1902. Die beiden Arten wurden in manchen Arbeiten
>von MIHELČIČ (z. B. 1953a, 1967a) auch als separate Taxa geführt.
>Verbr.: K, St, oT

?*Echiniscus simba* MARCUS, 1928
>Locus typicus: Feste Dürnstein: MARCUS 1928; N.
>Verbr.: N

Echiniscus spinulosus Doyère, 1840
> Verbr.: St, oT

Echiniscus tardus Mihelčič, 1951
> Locus typicus: fehlt in der Originalbeschreibung. Nachträglich wird der Fundort
> (Franz & Mihelčič 1954) als „Untertauern" ergänzt und danach (Mihelčič 1962a) als
> „Untertauern l. cl." definiert (St); siehe auch Mihelčič 1972.
> Verbr.: St
> Nomen dubium

Echiniscus testudo Doyère, 1840
> K, St, oT

Echiniscus trisetosus Cuénot, 1932
> Verbr.: St, K, S, oT

Echiniscus wendti Richters, 1903
> Verbr.: nT

Testechiniscus Kristensen, 1987

Testechiniscus spitsbergensis (Scourfield, 1897)
> Syn. *Echiniscus clavisetosus* Mihelčič, 1957. Lous typicus: Göltschach: Mihelčič 1957;
> K. (Synonymisierung: Dastych 1973); Mihelčič 1958.
> Syn. *Echiniscus rosaliae* Mihelčič, 1951. Locus typicus: „Studelgrat
> (Großglockner)...3250 m": Mihelčič 1951; oT. (Synonymisierung: Ramazzotti &
> Maucci 1983); Mihelčič 1953b.
> Syn. *Echiniscus spinuloides* Murray, 1907: Mihelčič 1953c.
> Verbr.: K, nT*, oT

Cornechiniscus Maucci & Ramazzotti, 1981

Cornechiniscus cornutus (Richters, 1906)
> Syn. *Cornechiniscus intermedius* Mihelčič, 1970. Locus typicus: nicht präzisiert: in der
> Originalbeschreibung das Taxon wurde von „Rosen- und Jauntal" (K) und „Osttirol"
> gemeldet: Mihelčič 1970. (Synonymisierung: Ramazzotti & Maucci 1983).
> Verbr.: B, K, N, St, oT

Pseudechiniscus Thulin, 1911

Pseudechiniscus conifer Richters, 1904
> Verbr.: oT

Pseudechiniscus dicrani Mihelčič, 1938
> Verbr.: St
> Nomen dubium

?Pseudechiniscus juanitae Barros, 1939
> Verbr.: N, S

Pseudechiniscus megacephalus MIHELČIČ, 1951
>*Pseudechiniscus megacephalus* nov. spec.: MIHELČIČ 1953b.
>Locus typicus: Rödschnitzer Moor: MIHELČIČ 1951; St.
>Verbr.: St, oT
>Nomen dubium

Pseudechiniscus pseudoconifer RAMAZZOTTI, 1943
>Verbr.: oT

Pseudechiniscus suillus (EHRENBERG, 1853)
>Verbr.: K, O, S, St, oT

Pseudechiniscus facettalis PETERSEN, 1951
>(= *Pseudechiniscus suillus facettalis*: MAUCCI 1974)
>Verbr.: O, S, oT

Pseudechiniscus victor (EHRENBERG, 1853)
>Verbr.: nT*
>Erstnachweis für Österreich.

Klasse Eutardigrada MARCUS, 1927

Ordnung Parachela SCHUSTER, NELSON, GRIGARICK et al., 1980

Familie Eohypsibiidae BERTOLANI & KRISTENSEN, 1904

Bertolanius ÖZDIKMEN, 2008

Bertolanius weglarskae (DASTYCH, 1972)
>Verbr.: nT*
>Erstnachweis für Österreich.

Familie Macrobiotidae THULIN, 1928

Dactylobiotus SCHUSTER, 1980

Dactylobiotus dispar (MURRAY, 1907)
>Verbr.: K, nT, oT, W

Dactylobiotus macronyx (DUJARDIN, 1851)
>Verbr.: N, O, St, oT
>Nomen dubium

Macrobiotus SCHULTZE, 1834

Macrobiotus crenulatus RICHTERS, 1904
>Verbr.: nT* (siehe unten)
>Erstnachweis für Österreich (s. u.).

?*Macrobiotus echinogenitus* RICHTERS, 1904
>Nach der Beschreibung von *M. dentatus* BINDA, 1974 (jetzt Syn. von *M. crenulatus*) und nachträglicher Synonymisierung dieser Art (siehe MAUCCI 1986, BINDA 1988), sollten die frühere Literaturangaben über *M. echinogenitus* sich meistens auf *M. crenulatus* RICHTERS beziehen (s.o.). Deshalb betreffen alle (?) Meldungen von MIHELČIČ über *M. echinogenitus* höchstwahrscheinlich *M. crenulatus*. Das Vorkommen *M. echinogenitus* (sensu BINDA 1988) im Bundesgebiet ist bestätigungsbedürftig.
>Verbr.: K, N, O, St, oT

Macrobiotus furcatus EHRENBERG, 1859
>Verbr.: St

Macrobiotus harmsworthi MURRAY, 1907
>Verbr.: K, N, O, S, St, oT
>(siehe Problematica)

Macrobiotus hufelandi SCHULTZE, 1834
>Verbr.: B, K, N, O, S, St, nT, oT
>(siehe Problematica)

Macrobiotus kolleri MIHELČIČ, 1951
>Locus typicus: „Oberlaussa, ca. 900 m": MIHELČIČ 1951; St.
>„Nächstverwandt mit *M. polaris* J. MURRAY": FRANZ & MIHELČIČ 1954.
>Verbr.: St, oT
>Nomen dubium

Macrobiotus macrocalix BERTOLANI & REBECCHI, 1993
>Verbr.: nT

Macrobiotus montanus MURRAY, 1910
>Verbr.: K, O, St, oT

?*Macrobiotus occidentalis* MURRAY, 1910
>Verbr.: oT

Macrobiotus pallari MAUCCI, 1954
>Verbr.: nT*
>Erstnachweis für Österreich.

Macrobiotus pseudohufelandi IHAROS, 1966
>Locus typicus: Ruster Hügelzug: IHAROS 1966; B.
>Verbr.: B

Macrobiotus ramoli DASTYCH, 2005
>Locus typicus: Hinterer Spiegelkogel, Ötztaler Alpen: DASTYCH 2005; nT.
>Verbr.: nT

?*Macrobiotus submorulatus* IHAROS, 1966
>Locus typicus: Kalenderberg bei Mödling: IHAROS 1966; N.
>Verbr. N

Macrobiotus striatus MIHELČIČ, 1949
>Locus typicus: fehlt in der Originalbeschreibung, nachträglich ergänzt MIHELČIČ (1962a) den Fundort als „Ennseck l. cl."; St.
>Verbr.: St
>Nomen dubium

Macrobiotus spectabilis Thulin, 1928
>Verbr.: nT*
>Erstnachweis für Österreich und die Alpen.

Macrobiotus terricola Mihelčič, 1949
>*Macrobiotus terricola* sp. n.: Mihelčič 1951
>*Macrobiotus terricola* nov. spec.: Mihelčič 1953b
>Locus typicus: fehlt in der Originalbeschreibung; nachträglich wird (Mihelčič 1962a)
>als „Ardning l. cl." ergänzt; St.
>Verbr.: St
>Nomen dubium

Paramacrobiotus Guidetti et al., 2009

Paramacrobiotus areolatus (Murray, 1907)
>„*Hypsibius* (*H.*) *areolatus*": Franz & Mihelčič 1954 (Strechengraben: St). Mihelčič
>(1972: 162) notiert über dem Fund „…die bei Franz 1954: 284 erwähnte Art *H.* (*H.*)
>*areolatus*… ist identisch mit *M. richtersi*…" (das korrekte Literaturzitat lautet „Franz
>& Mihelčič 1954" – die Bemerkung H.D.).
>Verbr.: O, St?, oT

Paramacrobiotus richtersi (Murray, 1913)
>Verbr.: B, K, N, O, S, St, oT
>(siehe Problematica)

Familie Murrayidae Guidetti, Rossi & Bertolani, 2005

Murrayon Bertolani & Pilato, 1988

Murrayon hibernicus (Murray, 1911)
>Verbr.: nT*
>Erstnachweis für Österreich.

Richtersius Pilato & Binda, 1989

Richtersius coronifer (Richters, 1903)
>= „*Macrobiotus islandicus* Richter" (sic!): in Franz & Mihelčič 1954; = „*Macrobiotus
>islandicus* Richters 1904" (sic!): Mihelčič 1962a. Nach Mihelčič (1972), eine
>Fehlbestimmung von *R. coronifer* aus Lochbach bei Lunz (N).
>Verbr.: K, N, St, oT, V

Minibiotus Schuster, 1980

Minibiotus intermedius (Plate, 1888)
>Verbr.: K, N, O, S, St, oT
>(siehe Problematica)

Familie Calohypsibiidae Pilato, 1969

Calohypsibius Thulin, 1928

Calohypsibius ornatus (Richters, 1900)
Verbr.: K, O, St, nT*, oT

Calohypsibius verrucosus (Richters, 1900)
Verbr.: K, St, oT

Familie Microhypsibiidae Pilato, 1998

Fractonotus Pilato, 1998

Fractonotus caelatus (Marcus, 1928)
Verbr.: nT*, S*
Erstnachweis für Österreich.

Familie Isohypsibiidae Marley, McInnes & Sands, 2011

Isohypsibius Thulin, 1928

Isohypsibius annulatus (Murray, 1905)
Verbr.: N

Isohypsibius austriacus (Iharos, 1966)
Locus typicus: Leitha-Gebirge bei Purbach: Iharos 1966; B.
Verbr.: B

Isohypsibius bellus (Mihelčič, 1971)
Locus typicus: „…im Süden der Ortschaft Amlach…zwischen der Drau und… dem Rauchkofel (1100 m)": Mihelčič 1971a.
Verbr.: oT
Nomen dubium

Isohypsibius belliformis (Mihelčič, 1971)
Locus typicus : Almach: Mihelčič 1971a; oT.
Verbr.: oT
Nomen dubium

Isohypsibius bulbifer (Mihelčič, 1957)
Locus typicus: in der Originalbeschreibung nicht präzisiert, das Material stammt aus „Maria Rain", „St. Johann" und „Rosental": (Mihelčič 1957). Nachträglich definiert Mihelčič (1962a) loc. typ. als „Maria Rain l. cl."; K.
Verbr.: K
Nomen dubium

Isohypsibius costatus (Mihelčič, 1971)
Locus typicus: „Amlach bei Lienz": Mihelčič 1971a; oT.
Verbr.: oT
Nomen dubium

Isohypsibius cyrilli (Mihelčič, 1949)
>*Hypsibius (Isohypsibius) cyrilli* n. sp.: Mihelčič 1951
>*Hypsibius (Isohypsibius) cyrilli* nov. spec.: Mihelčič 1953b
>Locus typicus: fehlt in der Originalbeschreibung; nachträglich ergänzt Mihelčič (1951) den Fundort als „Tellersack am Hochtor in den Gesäusealpen", „Holzgraben-Oberlaussa" und „Pürgschlachtermoor"; danach (Mihelčič 1962a) definiert loc. typ. als „Tellersack l. cl."; St.
>Verbr.: K, N, S, St, oT
>Nomen dubium

Isohypsibius dudichi (Iharos, 1964)
>Verbr.: S

Isohypsibius effusus (Mihelčič, 1971)
>Locus typicus nicht präzisiert: „schütterer Wälder (Kärnten) und Waldränder (Osttirol)": Mihelčič 1971b.
>Verbr.: K, oT.
>Nomen dubium

Isohypsibius franzi (Mihelčič, 1949)
>*Hypsibius (Isohypsibius) franzi* n. sp.: Mihelčič 1951
>*Hypsibius (Isohypsibius) franzi* nov. spec.: Mihelčič 1953b
>Locus typicus: fehlt in der Originalbeschreibung, nachträglich ergänzt Mihelčič (1962a) den Fundort als „Tristach l. cl."); oT.
>Verbr.: K, N, S, St, oT
>Nomen dubium

Isohypsibius granulifer Thulin, 1928
>Verbr.: nT, oT

Isohypsibius gulayi (Mihelčič, 1971)
>Locus typicus: nicht präzisiert, als Fundort wurde „Oberdrauburg" (K) und „Amlach bei Lienz" (oT) angegeben (Mihelčič 1971b).
>Verbr.: K, oT
>Nomen dubium

Isohypsibius hadzii (Mihelčič, 1938)
>Verbr.: oT
>Nomen dubium

Isohypsibius hypostomoides (Mihelčič, 1971)
>Locus typicus: Almach: Mihelčič 1971a; oT
>Verbr.: oT
>Nomen dubium

Isohypsibius leithaicus (Iharos, 1966)
>Locus typicus: Leitha-Gebirge bei Purbach: Iharos 1966; B.
>Verbr.: B

Isohypsibius lineatus (Mihelčič, 1969)
>Locus typicus: Michelbach bei St. Johann im Walde: Mihelčič 1969a; oT.
>Verbr.: oT
>Nomen dubium

Isohypsibius lunulatus (I<small>HAROS</small>, 1966)
      Verbr.: O, S

Isohypsibius mihelcici (I<small>HAROS</small>, 1964)
      „*H. mihelčiči* I<small>HAROS</small>": F<small>RANZ</small> 1975 (det. G. I<small>HAROS</small>).
      Verbr.: N

Isohypsibius nodosus (M<small>URRAY</small>, 1907)
      Verbr.: K, N, St, oT
      Nomen dubium

Isohypsibius pauper (M<small>IHELČIČ</small>, 1971)
      Locus typicus: Almach: M<small>IHELČIČ</small> 1971; oT.
      Verbr.: oT
      Nomen dubium

Isohypsibius prosostomus T<small>HULIN</small>, 1928
      Verbr.: K, N, O, S, St, nT, oT

Isohypsibius ronsisvallei B<small>INDA</small> & P<small>ILATO</small>, 1969
      Verbr.: S*
      Erstnachweis für Österreich.

Isohypsibius sattleri (R<small>ICHTERS</small>, 1902)
      Syn. *Isohypsibius bakonyiensis* (I<small>HAROS</small>, 1964).
      „*H. bakonyensis* (sic!) I<small>HAROS</small>": F<small>RANZ</small> 1975 (det. G. I<small>HAROS</small>).
      Verbr.: B, K, N, St, oT

?Isohypsibius schaudinni (R<small>ICHTERS</small>, 1909)
      „*H. schaudinni* Richt.": F<small>RANZ</small> 1975 (det. G. I<small>HAROS</small>).
      Verbr.: B, N, oT

Isohypsibius sellnicki (M<small>IHELČIČ</small>, 1962)
      Locus typicus: St. Johann im Walde: M<small>IHELČIČ</small> 1962b; oT.
      Verbr.: oT
      Nomen dubium

Isohypsibius solidus (M<small>IHELČIČ</small>, 1971)
      Locus typicus: „Kahlschlag bei Almach, Lienz": M<small>IHELČIČ</small> 1971a; oT.
      Verbr.: oT
      Nomen dubium

Isohypsibius tuberculoides (M<small>IHELČIČ</small>, 1949)
      Hypsibius (Isohypsibius) tuberculuides (sic!) n. sp.: M<small>IHELČIČ</small> 1951
      Hypsibius (Isohypsibius) tuberculoides n. sp: M<small>IHELČIČ</small> 1953b
      Hypsibius (Isohypsibius) tuberculoides nov. spec.: M<small>IHELČIČ</small> 1953b
      Locus typicus: fehlt in der Originalbeschreibung, nachträglich ergänzt M<small>IHELČIČ</small> (1951)
      den Fundort als „...gute Talwiese... (leg. Dr. Ing. F<small>RANZ</small>)"; danach präzisiert M<small>IHELČIČ</small>
      (1953a: 251) den Fundort als „Tristach"; oT. Die weiteren Fundorte siehe M<small>IHELČIČ</small>
      1972.
      Verbr.: O, St, oT
      Nomen dubium

Isohypsibius tortulosus (MIHELČIČ, 1959)
>Locus typicus: St. Johann im Walde: MIHELČIČ 1959; oT.
>Verbr.: oT
>Nomen dubium

Isohypsibius tuberculatus (PLATE, 1888)
>Verbr.: K, N, O, St, oT
>Nomen dubium

Isohypsibius undulatus THULIN, 1928
>Verbr.: St

Thalerius DASTYCH, 2009

Thalerius konradi DASTYCH, 2009
>Locus typicus: periglazial Zone bei Langtalerferner, Ötztaler Alpen:
>DASTYCH 2009b; nT.
>Verbr.: nT

Pseudobiotus NELSON, 1980

Pseudobiotus augusti (MURRAY, 1907)
>Verbr.: K, nT, oT, W

Familie Hypsibiidae PILATO, 1969

Hypsibius EHRENBERG, 1848

?Hypsibius convergens (URBANOWICZ, 1925)
>Verbr.: K, B, N, O, S, St, nT, oT
>(siehe Problematica)

Hypsibius dujardini (DOYÈRE, 1840)
>Verbr.: B, K, N, O, S, St, nT, oT

Hypsibius hypostomus BARTOŠ, 1935
>Verbr.: oT

Hypsibius klebelsbergi MIHELČIČ, 1959
>Locus typicus: „Gletscher im Tirolgebirge": MIHELČIČ 1959; nachträglich präzisiert
>MIHELČIČ (1962a) loc. typ. als „[Niederjochferner, Ötztaler Gletscher] l. cl."; nT.
>Pseudoendemit (DASTYCH 2009a); obligater Gletscherbewohner.
>Verbr.: nT

Hypsibius microps THULIN, 1928
>Verbr.: O, S, St, oT

Hypsibius pallidus THULIN, 1928
>Verbr.: K, S, St, oT

Hypsibius scabropygus CUÈNOT, 1929
>Syn. *Hypsibius callimerus* MARCUS, 1930: Locus typicus: „Wildspitze, 2830 m":
>MARCUS 1930; nT. (Synonymisierung: CUÈNOT 1932).
>Verbr.: K, nT

Mesocrista Pilato, 1987

Mesocrista spitzbergensis (Richters, 1903)
Verbr.: S*, nT*
Erstnachweis für Österreich.

Platicrista Pilato, 1987

Platicrista angustata (Murray, 1905)
Verbr.: K, O, oT

Diphascon Plate, 1888

(Untergattung *Adropion* Pilato, 1987)

Diphascon (A.) arduifrons Thuln, 1928
Verbr.: S, oT

Diphascon (A.) belgicae Richters, 1911
Verbr.: S, nT

Diphascon (A.?) marcusi (Rudescu, 1964)
Verbr.: oT
Nomen dubium

Diphascon (A.) prosirostre Thulin, 1928
Verbr.: O, S, oT

Diphascon (A.?) rivulare (Mihelčič, 1967)
Locus typicus: Almajurtal (oberes Lechtal): Mihelčič 1967c; nT.
Verbr.: K, nT
Nomen dubium

Diphascon (A.) scoticum Murray, 1905
Verbr.: B, K, N, O, S, St, nT*, oT
(siehe Problematica)

Diphascon (A.?) scoticum simplex (Mihelčič, 1971)
Hypsibius (Diphascon) scoticus J. Murray f. *simplex* Mihelčič, 1971: Mihelčič 1971a.
Locus typicus: „im Süden der Ortschaft Amlach…zwischen der Drau und… dem
Rauchkofel (1100 m)" (Mihelčič 1971a); oT.
Wahrscheinlich eine Art der Gattung *Itaquascon* Barros, 1939.
Verbr.: oT
Nomen dubium

Diphascon (A.?) speciosum (Mihelčič, 1971)
Locus typicus : „Amlach (Lienz)": Mihelčič 1971a; oT.
Verbr.: oT
Nomen dubium

(Untergattung *Diphascon* PLATE, 1888)

Diphascon (D.) alpinum MURRAY, 1906
>
> Verbr.: O, S, St, nT, oT
>
> Nomen dubium

Diphascon (D.?) bicorne (MIHELČIČ, 1971)
>
> *Hypsibius (Diphascon) scoticus* J. MURRAY f. *bicornis* MIHELČIČ 1971: MIHELČIČ 1971a.
> Locus typicus: „im Süden der Ortschaft Amlach…zwischen der Drau und… dem
> Rauchkofel (1100 m)": MIHELČIČ 1971a; oT.
> Verbr.: oT
> Nomen dubium

***Diphascon (D.) bullatum* MURRAY, 1905**
>
> Verbr.: K, B, St, oT

Diphascon (D.) chilenense PLATE, 1888
>
> Der taxonomische Status von *D. chilenense*: siehe DASTYCH 2003.
> Nomen dubium
> Verbr.: oT

Diphascon (D.?) elongatum (MIHELČIČ, 1959)
>
> Locus typicus: St. Johann im Walde: MIHELČIČ 1959; oT.
> Verbr.: oT
> Nomen dubium

Diphascon (D.?) gerdae (MIHELČIČ, 1951)
>
> Locus typicus: „unter dem Tristachersee": MIHELČIČ 1951; oT.
> Verbr.: oT
> Nomen dubium

Diphascon (D.?) latipes (MIHELČIČ, 1955)
>
> *Hypsibius (Diphascon) latipes* n. sp.: MIHELČIČ 1957
> Locus typicus: in der Originalbeschreibung: „Mauermoosen in Kärnten". Nachträglich
> MIHELČIČ (1962a) präzisiert loc. typ. als „Göltschach b. Maria Rain l. cl."; K.
> In beiden nachfolgenden Beschreibungen dieser Art unterscheiden sich die Abbildungen
> (s.o., MIHELČIČ 1955: 243, Fig. 1a–d und l. c. 1957: 669, Fig. 5a–d) deutlich in wichti-
> gen morphologischen Merkmalen!
> Verbr.: K
> Nomen dubium

Diphascon (D.?) mariae (MIHELČIČ, 1949)
>
> *Hypsibius (Diphascon) mariae* sp. n.: MIHELČIČ 1951
> *Hypsibius (Diphascon) mariae* nov. spec.: MIHELČIČ 1953b
> Locus typicus: fehlt in der Originalbeschreibung, nachträglich ergänzt MIHELČIČ (1951)
> den Fundort als „Osttirol und Steiermark", später (l. c. 1953a) werden „Dolomitas de
> Lienz", nämlich „Tristach" und „Kerschbaum", als auch Großglockner-Massiv mit
> „Heiligenblut" erwähnt und danach (l. c. 1953b) „Osttirol como en Estiria" genannt.
> Demnach (l. c. 1962a) wird loc. typ. als „Rabenstein l. cl." definiert (N). Weiterhin wur-
> de das Taxon auch von Moosham, Donnersbach und Donnersbachwald gemeldet (l. c.
> 1972).
> Verbr.: K, N, S
> Nomen dubium

Diphascon (D.?) nonbullatum (MIHELČIČ, 1951)
> Locus typicus: „unter dem Tristach bei Lienz": MIHELČIČ 1951; oT.
> Verbr.: oT
> Nomen dubium

Diphascon (D.) oculatum oculatum MURRAY, 1906
> Verbr.: O, nT*

Diphascon (D.) oculatum alpium MIHELČIČ, 1964
> Locus typicus: Zellersfeld bei Lienz: MIHELČIČ 1964; oT.
> Verbr.: oT
> Nomen dubium

Diphascon pingue (MARCUS, 1936)
> "*H. pinguis* MARCUS": FRANZ 1975 (det. G. IHAROS).
> Verbr.: N

Diphascon (D.) recamieri RICHTERS, 1911
> Fakultativer Gletscherbewohner.
> Verbr.: B, N, St, nT, oT

?*Diphascon (D.) trachydorsatum* BARTOŠ, 1937
> Verbr.: oT

Diphascon sp. 5
> *Hypsibius* (*Diphascon*) sp. 5: MIHELČIČ 1951, *Diphascon* sp. 5: MIHELČIČ 1953b.
> Fundort: Holzgraben-Oberlausa; oT.
> Verbr.: oT
> Nomen dubium

"*H. (D.) bipinguis*": MIHELČIČ 1955
> Nomen nudum

"*H. (D.) brachypus*": MIHELČIČ 1957: 654, Tafel 2.
> Nomen nudum

Astatumen PILATO, 1964

Astatumen trinacriae (ARCIDIACONO, 1962)
> Syn. *Itaquascon ramazzotti* IHAROS, 1966: Iharos 1966.
> "*Itaquascon ramazzotti* MAUCCI" (sic!): FRANZ 1975 (det. G. IHAROS).
> Verbr.: B, N, nT*

Astatumen bartosi (WEGLARSKA, 1958)
> Verbr.: S

Itaquascon BARROS, 1939

Itaquascon placophorum MAUCCI, 1973
> Verbr.: S

Itaquascon pawlowski WEGLARSKA, 1973
> Verbr.: S*
> Erstnachweis für Österreich.

Familie Ramazzottidae MARLEY, MCINNES & SANDS, 2011

Ramazzottius BINDA & PILATO, 1986

Ramazzottius agannae DASTYCH, 2011
>   Locus typicus: Zirmkogel, Ötztaler Alpen: DASTYCH 2011; nT.
>   Verbr.: nT

Ramazzottius cataphractus (MAUCCI, 1974)
>   Locus typicus: „FRANZ JOSEPH Haus am Großglockner Gletscher": MAUCCI 1974; K.
>   Verbr.: K

Ramazzottius montivagus (DASTYCH, 1983)
>   Verbr.: nT*
>   Erstnachweis für Österreich und die Alpen.

Ramazzottius oberhaeuseri (DOYÈRE, 1840)
>   Verbr.: K, N, St, oT

Hebesuncus PILATO, 1987

Hebesuncus conjungens (THULIN, 1911)
>   Verbr.: nT

Ordnung Apochela SCHUSTER, NELSON, GRIGARICK et al., 1980

Familie Milnesiidae RAMAZZOTTI, 1962

Milnesium DOYÉRE, 1840

Milnesium tardigradum DOYÈRE, 1840
>   Verbr.: K, N, O, S, St, oT, V
>   (siehe Problematica)

2. Problematica

Von allen 37 neuen Arten, die durch MIHELČIČ aus Österreich beschrieben worden sind, existiert kein Typenmaterial (!) – einzige Ausnahme ist *Hypsibius klebelsbergi*, der anhand vier Syntypen wieder beschrieben worden ist (DASTYCH 1993). Informationen über den locus typicus fehlen oft oder, was meist der Fall ist, er ist ungenau angegeben. Die diagnostischen Merkmale sind schlecht definiert, die Abbildungen sind unzureichend oder fehlen ganz. Daher sind wichtige Details nicht mehr nachprüfbar, d.h. alle 36 Namen sind keinem Taxon sicher zuzuordnen; sie werden im Folgenden daher als nomina dubia angesehen! Ein solches Vorgehen habe ich schon früher vorgeschlagen (l. c. 1993).

Ferner kommt hinzu, dass auch das Belegmaterial und weitere Angaben über sonstige durch MIHELČIČ gemeldete Arten fehlen. Auch für den Leser hilfreiche und überprüfbare taxonomische Kriterien oder solche Literaturhinweise in seinen

Publikationen bei der Auflistung von einzelnen Arten wurden fast nie angegeben. Außerdem repräsentieren viele von Mihelčič häufig gemeldeten Arten nach gegenwärtigen Erkenntnissen umfangreiche Komplexe von kryptischen Arten oder auch Taxa, die dringend der Revision bedürften. Dazu gehören z. B. *Bryodelphax parvulus, Macrobiotus hufelandi, M. harmsworthi, M. richtersi* (jetzt *Paramacrobiotus r.*), *Minibiotus intermedius, Hypsibius convergens, Diphascon scoticum, D. alpinum, Milnesium tardigradum* (siehe dazu u. a. BERTOLANI & REBECCHI 1994, BERTOLANI et al. 2011, GUIDETTI et al. 2009, CLAXTON 1998, DASTYCH 1988, PILATO 1974, MICHALCZYK et al. 2012). Aus diesen Gründen ist eine vollständige und kritische Überprüfung sämtlicher Mitteilungen von Mihelčič, die den größten Teil aller Angaben über die Bärtierchen ausmachen, unmöglich.

Die vorliegende Liste kann nicht auf die vielen taxonomischen Probleme eingehen, die mit der geschilderten Situation entstanden sind. Dennoch hoffe ich, dass sie als Grundlage für eine zukünftige Revision der Bärtierchen Österreichs dienen kann.

IV Literatur

BASSE, A. 1905: Beiträge zur Kenntnis des Baues der Tardigraden. — Zeitschr. wiss. Zool. **80**: 1–25.

BAUMANN, H. 1922: Die Anabiose der Tardigraden. — Zool. Jb. Syst. Ökol. Geogr. Tiere **45**: 501–556.

BERTOLANI, R. & REBECCHI, L. 1993. A revision of the *Macrobiotus hufelandi* group (Tardigrada, Macrobiotidae), with some observations on the taxonomic characters of eutardigrades. — Zool. Scr. **22** (2): 127–152.

BERTOLANI, R., REBECCHI, L., GIOVANNINI, I., & CESARI, M. 2011: DNA barcoding and integrative taxonomy of *Macrobiotus hufelandi* C.A.S. SCHULTZE 1834, the first tardigrade species to be described, and some related species. — Zootaxa **2997**: 19–36.

BINDA M. G. 1974: Tardigradi della Valtellina. — Animalia, Catania l (1–3): 201–216.

BINDA, M.G. 1988: Ridiscrizione di *Macrobiotus echinogenitus* RICHTERS, 1904 e sul valore di buona specie di *Macrobiotus crenulatus* RICHTERS, 1904 (Eutardigrada). — Animalia, Catania **15** (1/3): 201–210.

CLAXTON, S.K. 1998: A revision of the genus *Minibiotus* (Tardigrada, Macrobiotidae) with descriptions of eleven new species from Australia. — Rec. Austral. Mus. **50**: 125–160.

DABERT, M., DASTYCH, H., HOHBERG, K. & DABERT, J. 2014: Phylogenetic position of the enigmatic clawless eutardigrade genus *Apodibius* DASTYCH, 1983 (Tardigrada), based on 18S and 28S rRNA sequence data from its type species *A. confusus*. — Molec. Phylogen. Evol. **70**: 70–75.

DASTYCH, H. 1988: The Tardigrada of Poland. — Monografie Fauny Polski, Warszawa-Krakow **16**: 1–255.

DASTYCH, H. 1993: Redescription of the cryoconital tardigrade *Hypsibius klebelsbergi* MIHELČIČ, 1959, with notes on the microslide collection of the late Dr. F. MIHELČIČ. — Veröff. Mus. Ferdinandeum, Innsbruck **73**: 5–12.

DASTYCH, H. 2003: *Diphascon langhovdense* (SUDZUKI, 1964) stat. nov., a new taxonomic status for the semi-terrestrial tardigrade (Tardigrada). — Acta Biol. Benrodis, Düsseldorf **12**: 19–25.

DASTYCH, H. 2005: *Macrobiotus ramoli* sp. nov., a new tardigrade species from the nival zone of the Ötztal Alps, Austria (Tardigrada). — Mitt. Hamburg. Zool. Mus. Inst. **102**: 21–35.

DASTYCH, H. 2009a: Tardigrada (Bärtierchen). — In RABITSCH, W. & ESSL, F. (Hrsg.): Endemiten – Kostbarkeiten in Österreichs Pflanzen- und Tierwelt. — Naturwissenschaftlicher Verein für Kärnten und Umweltbundesamt GmbH, Klagenfurt & Wien, S. 383–384.

DASTYCH, H. 2009b: *Thalerius konradi* gen. nov., sp. nov., a new tardigrade from the periglacial area of the Ötztal Alps, Austria (Tardigrada). — Contr. Nat. Hist., Bern **12**: 391–402.

DASTYCH, H. 2011: *Ramazzottius agannae* sp. nov., a new tardigrade species from the nival zone of the Austrian Central Alps (Tardigrada). — Ent. Mitt. Zool. Mus. Hamburg **15**: 237–253.

DASTYCH, H., KRAUS, H. & THALER, K. 2003: Redescription and notes on the biology of the glacier tardigrade *Hypsibius klebelsbergi* MIHELČIČ, 1959 (Tardigrada), based on material from the Ötztal Alps, Austria. — Mitt. Hamburg. Zool. Mus. Instit. **100**: 77–100.

DASTYCH, H. & THALER, K. 2002: The tardigrade *Hebesuncus conjungens* (THULIN, 1911) in the Alps, with notes on morphology and distribution (Tardigrada). — Ent. Mitt. Zool. Mus. Hamburg **14** (166): 83–94.

EHRENBERG, Ch.G. 1853: Diagnosis novarum formarum. — Monatsberichte der Königlichen Preussische Akademie des Wissenschaften zu Berlin, 526–533.

FRANCÉ, R.H. 1921: Das Edaphon. Untersuchungen zur Oekologie der bodenbewohnenden Mikroorganismen. — In: Arbeiten aus dem Biologischen Institut München, Franckh'sche Verlagshandlung, Stuttgart, 2, 1–99.

FRANZ, H. 1950: Bodenzoologie als Grundlage der Bodenpflege (mit besonderer Berücksichtigung der Bodenfauna in den Ostalpen und im Donaubecken). — Akademie-Verlag Berlin, 316 pp.

FRANZ, H. 1975: Die Bodenfauna der Erde in biozönotischer Betrachtung. — Wiesbaden: Franz Steiner Verlag, 766 pp.

FRANZ, H. 1983: Die Nordost-Alpen im Spiegel ihrer Landtierwelt. a) Die Bärtierchen – eine im Bezirk schlecht erforschte Tiergruppe. — In RESSL, F. (Hrsg.): Die Tierwelt des Bezirkes Scheibbs. — Scheibbs **2**: 74–76.

FRANZ, H. & MIHELČIČ, F. 1954: Ordnung: Tardigrada. — In Franz H. (Hrsg.): Die Nordost-Alpen im Spiegel ihres Landtierwelt. — Innsbruck **1**: 281–287.

GREVEN, H. 2013: 2.Tardigrada, Bärtierchen. — In WESTHEIDE, W. & RIEGER, G. (Hrsg.): Spezielle Zoologie, Teil 1: Einzeller und Wirbellose Tiere, 3. Auflage. Berlin-Heidelberg: Springer Verlag.

GUIDETTI, R., SCHILL, R.O., BERTOLANI, R., DANDEKAR, T. & WOLF, M. 2009: New molecular data for tardigrade phylogeny, with the erection of Paramacrobiotus gen. nov. — J. Zool. Syst. Evol. Res. **47**: 315–321.

GUIDETTI, R. & BERTOLANI, R. 2011: Phylum Tardigrada DOYÈRE, 1840. — Zootaxa **3148**: 96–97.

HOSCHITZ, M. 2004: Moss-living nematodes from an Alpine summit (Dachstein, Austria). — Verh. Zool.-Bot. Gesellschaft Österreich, Wien **140**: 93–98.

IHAROS G. 1966: Beiträge zur Kenntnis der Tardigraden-Fauna Österreichs. — Acta Zool. Hungar. **12** (1–2): 123–127.

JANETSCHEK, H. 1957. Die Tierwelt des Raumes von Kufstein. — Schlern-Schriften, Innsbruck **156**: 203–275.

KIEHL, E., DASTYCH, H., D'HAESE, J., & GREVEN, H. 2007. The 18S rDNA sequences support polyphyly of the Hypsibiidae (Eutardigrada). — Proc. 10th Internat. Symp. Tardigrada (PILATO, G. & REBECCHI, L. Eds.). — J. Limn. **66**: 21–25.

KOFLER, A. 1978. Biographie und Bibliographie des Acarinologen und Tardigradologen Pfarrer Dr. Franz MIHELČIČ (1898–1977). — Ber. naturwiss.-med. Ver. Innsbruck **65**: 213–224.

KRAUS, H. 1977: *Hypsibius* (*Hypsibius*) *klebelsbergi* MIHELČIČ, 1959 (Tardigrada) aus dem Kryokonit des Rotmoosferner (Ötztaler Alpen). — Institut für Zoologie der Universität Innsbruck, Dissertation, 189 pp. (nicht publiziert).

KÜHN, G. 1940: Zur Ökologie und Biologie der Gewässer (Quellen und Abflüsse) des Wassergspreng bei Wien. — Arch. Hydrobiol. **36**: 157–192.

MARCUS, E. 1928: Bärtierchen (Tardigrada). – In: Spinnentiere oder Arachnoidea. Die Tierwelt Deutschlands und der angrenzenden Meeresteile. — Jena **12** Teil, 4: 1–230.

MARCUS, E. 1929: Tardigrada. – BRONN's Klassen und Ordnungen des Tierreichs. — Leipzig **5** (Abt. 4): 1–608.

MARCUS, E. 1930: Beiträge zur Tardigradensystematik. — Zool. Jb. Syst. Ökol. Geogr. Jena **59**: 363–386.

MARCUS, E. 1936: Tardigrada. — Das Tierreich **66**: 1–340.

MAUCCI W. 1974: *Hypsibius* (*H.*) *cataphractus* (Tardigrada: Macrobiotidae) und weitere Nachrichten über Tardigraden aus Österreich. — Ber. naturwiss. med. Ver. Innsbruck **61**: 83–86.

MAUCCI, W. 1986: Tardigrada. In: Fauna d' Italia. — Bologna: Calderini **24**: 388 pp.

MAYER, G., MARTIN, C., RÜDIGER, J., KAUSCHKE, S., STEVENSON, P.A., POPRAWA, I., HOHBERG, K., SCHILL, R. O., PFLÜGER, H.-J. & SCHLEGEL, M. 2013: Selective neuronal staining in tardigrades and onychophorans provides insights into the evolution of segmental ganglia in panarthropods. — BMC Evolutionary Biology **13** (230): 1–16.

McInnes, S.J. 1994: Zoogeographic distribution of terrestrial/freshwater tardigrades from current literature. — J. Nat. Hist. **28**: 257–352.

Michalczyk, L., Welnicz, W., Frohme, M. & Kaczmarek, L. 2012: Redescriptions of three *Milnesium* Doyère, 1840 taxa (Tardigrada: Eutardigrada: Milnesiidae), including the nominal species for the genus. — Zootaxa **3154**: 1–20.

Micoletzki, H. 1910/11: Zur Kenntnis des Faistenauer Hintersees bei Salzburg, mit besonderer Berücksichtigung faunistischer und fischereilicher Verhältnisse. — Intern. Rev. Hydrobiol. Hydrogr. **3**: 506–542, Tafel V–III.

Mihelčič, F. 1939: Beitrag zur Kenntnis der Tardigrada Jugoslaviens. 1. — Prorodoslovne Razprave, Ljubljana **3** (12): 331–336.

Mihelčič, F. 1949: Nuevos biotopos de Tardigrados. Contribucion al conocimiento de la ecologia de los Tardigrados. — Anal. Edafol. Fisiol. Veget., Madrid **8**: 511–526.

Mihelčič, F. 1951: Beitrag zur Systematik der Tardigraden. — Arch. Zool. Ital., Torino **36**: 57–102.

Mihelčič, F. 1952: Contribucion al estudio de la ecologia de los Tardigrados que habitan suelos de humus. — Anal. Edafol. Fisiol. Veget., Madrid **11**: 407–446

Mihelčič, F. 1953a. Contribucion al conocimento de los Tardigrados con especial consideracion de los Tardigrados de Osttirol. — Anal. Edafol. Fisiol. Veget., Madrid **12** (5): 243–274.

Mihelčič, F. 1953b: Contribucion al conocimento de los Tardigrados con especial consideracion de los Tardigrados de Osttirol (II). — Anal. Edafol. Fisiol. Veget., Madrid **12** (5): 431–479.

Mihelčič, F. 1953c: Vorläufiger Bericht über die in den Wäldern um Göltschach (Maria Rain, Kärnten) festgestellten Tardigraden und Nematoden. — Carinthia II, **63 /143**: 115–117.

Mihelčič, F. 1955: Zur Ökologie und Verbreitung der Gattung *Hypsibius* (Tardigrada). — Bonner Zool. Beitr. **6**: 240–244.

Mihelčič, F. 1957: Contribucion a la ecologia de los Tardigrados de suelos humedos. — Anal. Edafol. Fisiol. Veget., Madrid **16** (5): 651–671.

Mihelčič, F. 1958: Eine neue Tardigradenart feuchter Standorte. — Zool. Anz., Jena **160**: 105–108.

Mihelčič, F. 1959: Zwei neue Tardigraden aus der Gattung *Hypsibius* Thulin aus Osttirol (Österreich). Systematisches zur Gattung *Hypsibius* Thulin. — Zool. Anz., Jena **163**: 254–261.

Mihelčič, F. 1962a: Tardigrada. — In Strouhal, H. (Ed.): Catalogus Faunae Austriae, Teil VI: 6–11. — Wien: Österr. Akad. Wiss.

Mihelčič, F. 1962b: Eine neue *Hypsibius* (*Isohypsibius*) Art (Tardigrada) aus Osttirol (Tirol). — Zool. Anz., Jena **168**: 239–241.

Mihelčič, F. 1963a: Bärtierchen, die auf Gletschern leben. — Mikrokosmos, Stuttgart **52** (2): 44–46.

Mihelčič, F. 1963b: Können Tardigraden im Boden leben? — Pedobiol. **2**: 96–101.

MIHELČIČ, F. 1964: Tardigraden einiger Felsenmosse in Osttirol. — Verh. Zool.-Bot. Ges. Wien **103/104**: 94–100.

MIHELČIČ, F. 1965a: Tardigraden einiger Auwälder in Osttirol. — Veröffentlichungen aus dem Haus der Natur, Salzburg, **7** (2): 1–8 (Seiten nicht nummeriert).

MIHELČIČ, F. 1965b: Schneetälchen als Lebensstäten für Tardigraden. — Veröffentlichungen aus dem Haus der Natur, Salzburg **7** (2): 1–7 (Seiten nicht nummeriert).

MIHELČIČ, F. 1965c: Zur Kenntnis der Entwicklung der Tardigradenzönosen während der Verrottung der Streu. — Zool. Anz., Jena **174** (2): 150–156.

MIHELČIČ, F. 1967a: Ein Beitrag zur Kenntnis der Tardigrada der Steiermark. — Mitt. Naturwiss. Ver. Steiermark **97**: 67–76.

MIHELČIČ, F. 1967b: Baummoose und Flechten als Lebensstäten für Tardigraden. — Carinthia II, **77/157**: 227–236.

MIHELČIČ, F. 1967c: Ein Beitrag zur Kenntnis der Süsswassertardigraden Kärntens. — Carinthia II, **77/157**: 222–226.

MIHELČIČ, F. 1967d: Der Boden als Wohnraum für Tardigraden. — An. Edaf. Fisiol. Veget., Madrid **26** (1–4): 145–157.

MIHELČIČ, F. 1969a: Zur Kenntnis der Tardigraden Osttirols. — Veröff. Mus. Ferdinandeum, Innsbruck **49**: 113–130.

MIHELČIČ, F. 1969b: Ein Beitrag zur Kenntnis der Tardigraden des Göltschacher Moores bei Maria Rain (Kärnten). — Carinthia II, **79/159** :138–144.

MIHELČIČ, F. 1970: Beiträg zur Gattung *Pseudechiniscus* THULIN (Phylum Tardigrada). — Carinthia II, **160/80**: 105–110.

MIHELČIČ, F. 1971a: Beobachtungen an Tardigraden Osttirols (I). — Veröff. Tiroler Landesmus. Ferdinandeum, Innsbruck **51**: 119–140.

MIHELČIČ, F. 1971b: Kurzbericht über Tardigraden einiger Böden Kärntens. — Carinthia II, **161/81**: 75–85.

MIHELČIČ, F. 1972: Zur Kenntnis der Tardigraden der Steiermark. — Mitt. Naturwiss. Ver. Steiermark **102**: 157–167.

PILATO, G. 1974: Studio su *Diphascon scoticum* J. MURR., 1905 (Eutardigrada) e alcune specie ritenute ad esso affini. — Animalia, Catania **1** (1/3): 73–88.

PILATO, G. 1987: Revision of the genus *Diphascon* PLATE, 1889, with remarks on the subfamily Itaquasconinae. — In R. Bertolani (Ed.): Biology of Tardigrades. — Proc. 4th Int. Symp. Tardigrada, Modena, September 3–5, 1985. Selected Symposia and Mongraphs U.Z.I., Mucchi, Modena **1**: 337–357.

PILATO, G. & BINDA, M. G. 2010: Definition of families, subfamilies, genera and subgenera of the Eutardigrada, and keys to their identification. — Zootaxa **2404**: 1–54.

PLATE, L. 1888: Beiträge zur Naturgeschichte der Tardigraden. — Zool. Jb. Anat. Ontog. Tiere, Jena **3**: 487–550.

Post, B. 2012: Structural and functional interactions in Alpine cryoconite holes with special emphasis on Tardigrades. — Universität Innsbruck, Masterarbeit (MSc), 72 pp. (nicht publiziert).

Ramazzotti, G. & Maucci, W. 1983. Il phylum Tardigrada (III edizione riveduta e aggiornata). — Mem. Istit. Ital. Idrobiol., Pallanza **41**: 1–1012.

Schuetz, G. 1987: A one-year study on the population dynamics of *Milnesium tardigradum* Doyère in the lichen *Xanthoria parietina* (L.) Th. Fr. — In Bertolani, R. (Ed.): Biology of Tardigrades. — Proc. 4th Int. Symp. Tardigrada, Modena, September 3–5, 1985. Selected Symposia and Mongraphs U.Z.I., Mucchi, Modena **1**: 217–228.

Steinböck, O. 1957: Über die Fauna der Kryokonitlöcher alpiner Gletscher. — Der Schlern, Bozen **31**: 65–70.

Thaler, K. 1999: Nival invertebrate animals in the East Alps: a faunistic overview (2.5. Tardigrada). – In Margesin, R. & Schinner, F. (eds.): Cold-adapted organisms. — Heidelberg, New York: Springer-Verlag, pp. 165–179.

Tilzer, M. 1968: Zur Ökologie und Besiedlung des hochalpinen hyporheischen Interstitial im Arlberggebiet. — Archiv für Hydrobiologie, Stuttgart **65**: 253–308.

Vornatscher, J. 1938: Faunistische Untersuchungen des Lusthauswassers im Wiener Prater. — Intern. Rev. Hydrobiol. Hydrogr., Leipzig **37**: 320–363.

Wright, J.C. 2001: Cryptobiosis 300 years on from van Leuwenhoek: what we have learned about tardigrades? — Zool. Anz., Jena **240**: 563–582.

Adresse des Autors:

Dr. Hieronymus Dastych
Zoologisches Museum, Universität Hamburg
Martin-Luther-King-Platz 3, D-20146 Hamburg, Deutschland
Email: dastych@zoologie.uni-hamburg.de
http://www.uni-hamburg.de/biologie/BioZ/zmh/ent/dastych.html

Odonata (Insecta)

Werner E. Holzinger, Andreas Chovanec
& Johann A. Waringer

Summary: This checklist summarizes our present knowledge of the Austrian Odonata species inventory. The Austrian fauna is diverse, combining Alpine and Mediterranean elements. Currently, 78 Odonata species are known from Austria. The paper provides updates on distribution and contains introductory chapters dealing with Odonata biology and bioindication, a brief history on Odonata research, and extensive references.

Zusammenfassung: Diese Arbeit stellt eine Aktualisierung des kommentierten Verzeichnisses der Libellen Österreichs dar. Die sehr diverse österreichische Libellenfauna umfasst neben alpinen auch mediterrane Elemente; aktuell sind 78 Arten aus Österreich nachgewiesen. Neben neuen ergänzenden Angaben für die einzelnen Taxa sind auch ein einleitendes Kapitel über die Biologie der Odonaten, ein Überblick über Erforschungsgeschichte und Bioindikation sowie ein ausführliches Literaturverzeichnis enthalten.

Key words: Odonata, Austria, checklist, updates, distribution, biology, research history, bioindication references

Werner E. Holzinger et al.

I Einleitung

Libellen stellen eine stammesgeschichtlich alte Insektengruppe dar. Die fossile Ordnung der Protodonata ist seit dem Oberkarbon (vor ca. 325 Millionen Jahren) durch zahlreiche Fossilbelege gut dokumentiert (CORBET 1999). Einen grossen Bekanntheitsgrad erreichte die Art *Meganeura monyi* BROGNIART aus den Kohlenflözen von Commentry in Südostfrankreich, da sie mit 670 mm Flügelspannweite das größte europäische Insekt ist.

Von weltweit etwa 6.000 Arten aus 30 Familien kommen in Europa 135 Arten (10 Familien) und in Österreich 78 Arten aus 9 Familien vor. Sie zählen zu den größten, auffälligsten und bekanntesten Insekten Österreichs. Mit dieser Arbeit wird eine Aktualisierung des kommentierten Verzeichnisses der Libellen Österreichs vorgelegt; zudem wird ein summarischer Überblick über die Erforschungsgeschichte der Libellenfauna Österreichs, die Biologie der Odonaten und über ihre Bedeutung als Bioindikatoren präsentiert.

II Allgemeiner Teil

1. Erforschungsgeschichte und aktueller Kenntnisstand

Seit den Anfängen entomologisch-faunistischer Forschungen in Österreich sind Libellen als große und auffällige Insekten im Fokus der Insektenforscher und -innen. Bereits in Nikolaus Poda von Neuhaus' „Insecta Musei Gracensis" (PODA 1761) werden acht Arten für die Umgebung von Graz genannt. Viele weitere Faunisten – insbesondere Carl AUSSERER (Innsbruck), Christian Casimir BRITTINGER (Steyr), Friedrich Moritz BRAUER (Wien), Carl Wilhelm DALLA TORRE VON TURNBERG-STERNHOF (Innsbruck), Gerald MAYER (Linz), Fritz PRENN (Kufstein), Anton SCHWAIGHOFER (Graz), Roman PUSCHNIG (Klagenfurt), Franz Josef Maria WERNER (Wien), Hans STROUHAL (Wien), Wolfgang PICHLER (Leoben) und Roman PUSCHNIG (Klagenfurt) – trugen dazu bei, dass Douglas St. Quentin (Wien) im ersten Catalogus Faunae Austriae, der sich den Libellen widmete (ST. QUENTIN 1959), bereits 75 Arten für Österreich nennen konnte. Im dritten Viertel des 20. Jahrhunderts widmeten sich unter anderem Walter CHARWAT (Salzburg), Harald HEIDEMANN (Bruchsal, D), Alois KOFLER (Lienz), Franz RESSL (Purgstall), Wilfried STARK (Graz) und Günter THEISCHINGER (Linz; heute Lidcombe, Australien) intensiv den Libellen Österreichs, sodass die im Jahr 1976 aktualisierte Checkliste der Libellen Österreichs von Martin LÖDL (Wien) bereits 79 Arten umfasste (LÖDL 1976a, b).

Spätere Autoren – Armin LANDMANN (Innsbruck), Gerhard LEHMANN (Kufstein), Andrea WARINGER-LÖSCHENKOHL (Wien) und Johann WARINGER (Wien) seit den 1980er-Jahren, Helwig BRUNNER (Graz), Andreas CHOVANEC (Wien), Hans EHMANN (Werfen), Werner E. HOLZINGER (Graz), Kurt HOSTETTLER (Romanshorn, CH), Gerold LAISTER (Linz), Eva SCHWEIGER-CHWALA (Mauerbach) und Rainer RAAB (Deutsch-Wagram) seit den 1990ern sowie rezent zudem Thomas FRIESS (Graz), Patrick GROS (Salzburg),

Helmut Höttinger (Wien), Brigitte Komposch (Graz), Franz Mungenast (Imst), Werner Petutschnig (Klagenfurt), Otto Samwald (Fürstenfeld), Maria Schwarz-Waubke (Kirchschlag), Maria Schindler (Wien), Wolfgang Schweighofer (Artstetten), Hermann Sonntag (Wattens), Friedrich & Margit Stich (Ferlach), Siegfried Wagner (Landskron) und viele weitere – trugen zahlreiche Daten zusammen, die in eine monographische Darstellung der Libellenfauna Österreichs mündeten (Raab et al. 2007). Dieser ist auch eine umfangreichere Darstellung der Erforschungsgeschichte der Libellenfauna Österreichs zu entnehmen. Eine Bibliographie zur Libellenfauna Österreichs legte Rainer Raab (1994) vor, Landesfaunen existieren für Kärnten (Holzinger & Komposch 2012), die Steiermark und das Burgenland (Stark 1976), Niederösterreich (Raab & Chwala 1997), Oberösterreich (Laister 1996a, b), Tirol (Landmann et al. 2005) und Vorarlberg (Hostettler 2001), für Salzburg ist eine solche in Vorbereitung (Gros & Ehmann, pers. comm.).

Aktuell liegen für 78 Libellenarten aus 25 Gattungen, die vier Familien der Kleinlibellen (Zygoptera) und fünf Familien der Grosslibellen (Anisoptera) zuzurechnen sind, sichere Nachweise aus Österreich vor. 46 Arten sind in allen Bundesländern nachgewiesen. Die Artenzahlen der einzelnen Bundesländer zeigt Abbildung 1.

Mit *Coenagrion hylas* ssp. *freyi* ist eine Unterart als Subendemit Österreichs anzusehen (vgl. Chovanec 2009), deren außerösterreichische Vorkommen (Bayern) allerdings erloschen sind.

Insgesamt ist damit das Spektrum autochthoner Arten und deren regionaler Verbreitung sehr gut bekannt, Nachweise weiterer Arten sind nur für migrierende Taxa (z. B. *Boyeria irene* (Fonscolombe, 1838)), verschleppte Arten (z. B. Kipping 2006) und bei Arealausweitungen zu erwarten. Aufgrund der Gefährdung ihrer Lebensräume ist – ohne rasche Gegenmassnahmen zum Schutz der Arten – zumindest bei *Leucorrhinia albifrons*, *Leucorrhinia caudalis* und *Sympetrum depressiusculum* kurz- bis mittelfristig mit dem endgültigen Erlöschen der Vorkommen in Österreich zu rechnen.

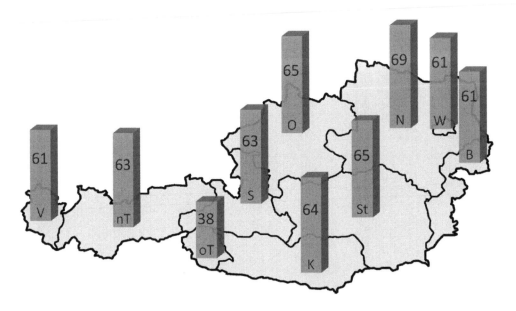

Abb. 1: Anzahl der aus den Österreichischen Bundesländern nachgewiesenen Libellenarten.
Anmerkung: Nord- und Osttirol werden getrennt dargestellt.

2. Biologie der Libellen

Aufgrund des zeitlichen Auftretens der Imagines im Jahreslauf lassen sich die einheimischen Arten in die drei phänologischen Gruppen der Winterlibellen, der Frühjahrs- und der Sommerarten einteilen (SCHMIDT 1985). Die Winterlibellen der Gattung *Sympecma (S. fusca, S. paedisca)* sind die einzigen einheimischen Libellen, die regulär als Imagines überwintern und daher mit über 10 Monaten die höchste imaginale Lebensspanne unserer Arten besitzen. Die Adulttiere überdauern den Hochwinter in Form einer Winterruhe, wobei die Tiere von abdominalen Fettreserven zehren und unbeweglich entweder frei auf Vegetationsstrukturen sitzend oder in Moospolstern, Heidekraut oder unter Steinen verborgen angetroffen werden (JÖDICKE 1997).

Zur phänologischen Gruppe der Frühjahrsarten gehören u. a. *Calopteryx virgo*, die Coenagrioniden *Coenagrion lunulatum, Erythromma najas* und *Pyrrhosoma nymphula*, weiters *Cordulegaster boltonii, Anax imperator*, die Libelluliden *Leucorrhinia rubicunda, Libellula depressa* und *Orthetrum cancellatum* und die Corduliide *Somatochlora metallica*. Diesen Arten ist gemeinsam, dass die Imagines von der 1. Maidekade bis zur 1. Junidekade schlüpfen und dass die volle Reproduktionsperiode etwa in den Zeitraum 3. Maidekade bis 2. Julidekade fällt.

Zu den Sommerarten zählen schliesslich jene Arten, die von der 2. Junidekade bis in die 1. Augustdekade schlüpfen und deren Hauptfortpflanzungszeit von der

2. Julidekade bis in den September beobachtet werden kann. Bekannte Beispiele für diese Hochsommerlibellen finden sich u. a. unter den Aeshnidae (z. B. *Aeshna cyanea, A. grandis, A. mixta*), den Libellulidae (z. B. *Sympetrum danae, S. vulgatum, S. striolatum*), den Lestidae (z. B. *Lestes dryas, L. sponsa, L. viridis*) (CORBET 1999; WILDERMUTH 1994) und unter den Coenagrionidae (*Ischnura elegans, Coenagrion puella, C. pulchellum, Enallagma cyathigerum*; PARR 1970).

Die meisten einheimischen Libellenarten sind an wolkenlosen, windstillen, sonnigen Tagen am aktivsten. Beobachtungen an *Coenagrion puella* verdeutlichen den kombinierten Einfluss von Windgeschwindigkeit und Bewölkungsgrad auf die Flug- und Eiablageaktivität: die Windtoleranz sinkt mit zunehmendem Bewölkungsgrad. Aber auch bei optimalem Wetter beginnt die Flugaktivität der Hufeisenazurjungfer im August kaum vor 9 Uhr Sonnenzeit, erreicht zur wärmsten Zeit um Mittag ihren Höhepunkt und lässt ab 15 Uhr bereits deutlich nach (WARINGER 2007). Neben diesem unimodalen, mittagszentrierten Muster finden sich viele Belege für eine Ausdehnung der Flugaktivität heimischer Arten in die Dämmerung hinein. Beispiele hierfür gibt es vor allem bei den Corduliidae (*Cordulia aenea, Somatochlora metallica*), bei *Aeshna*-Arten (*A. cyanea, A. grandis, A. viridis*) sowie bei *Anax imperator*.

Zum Schlüpfen verlassen die Libellenlarven ihr Wohngewässer meist am Abend. Als artspezifische Schlüpfsubstrate dienen Uferabbrüche und Kieselsteine an Flussufern, emerse Wasserpflanzenteile, Röhrichtpflanzen und Hochstauden im Verlandungsbereich sowie Wurzelbärte, Stämme und Äste der Ufersträucher und -bäume. Sind im Umfeld der Wasseranschlagslinie keine geeigneten Strukturen vorhanden, werden oft beträchtliche Entfernungen über Land zurückgelegt. So konnte MÜLLER (1999) Exuvien von *Cordulegaster bidentata* bis zu 12 m vom Bachufer entfernt und bis zu 3,7 m hoch an Baumstämmen feststellen.

Der erste längere Flug der frischgeschlüpften, noch unausgefärbten Libellen führt sie weg vom bisherigen Wohngewässer, und die Zeit bis zur Reproduktionsphase wird oft in beträchtlicher Entfernung von aquatischen Biotopen in Wäldern, Wiesen oder Gebüschen verbracht. Nach Vollendung der Maturationsperiode kehren die voll geschlechtsreifen Tiere zum Gewässer zurück, wobei manche Arten, wie z. B. *Lestes barbarus*, einen hohen Bindungsgrad an ihr ursprüngliches Brutgewässer aufweisen (UTZERI et al. 1976).

Libellen sind vollendete Flugjäger und zu diesem Zweck morphologisch optimal ausgestattet, wobei die bedornten Beine durch den schräg gestellten Pterothorax einen funktionalen, nach vorn ausgerichteten Fangkorb bilden. Ihre hocheffektive Flugmuskulatur umfasst bis zu 63 % des Totalkörpergewichts (CORBET 1999). Die Flugjäger (z. B. Aeshnidae und Corduliidae) führen bei mäßiger Geschwindigkeit ausgedehnte Jagdflüge um Baumkronen und entlang von Hecken und Waldsäumen aus, wobei es zwischen gemeinsam jagenden Arten kaum Interaktionen gibt (KAISER 1974). Ansitzjäger (z. B. Libellulidae, Zygoptera) besetzen hingegen Sitzwarten, die gegen Konkurrenten verteidigt werden (GORB 1994, MOORE 1991), überraschen ihre Beute durch überfallsartige Jagdflüge (PARR 1983, REHFELDT et al. 1993) und kehren nach wenigen Sekunden zur Warte zurück. Zusätzlich zu fliegenden Insekten werden ungeflügelte Blattläuse, in Spinnennetzen hängende Insekten sowie Spinnen direkt von der Unterlage abgesammelt. Ist die Beute erfasst, wird sie von den Flugjägern meist direkt im Flug verzehrt und ungenießbare Körperteile fallengelassen (MARTENS & WIMMER 1996). Das Nahrungsspektrum adulter Libellen umfasst Chironomidae, Culicidae und andere Dipteren, Libellen, Schmetterlinge, Ameisen und Bienen (SUKHACHEVA 1996).

Die genitalmorphologischen Besonderheiten der Odonaten (Ausbildung eines akzessorischen, sekundären männlichen Kopulationsapparates) beeinflussen die Paarungssequenz in einzigartiger Weise. Da keine direkte innere Verbindung der Keimdrüsen zum funktionellen Penis besteht, müssen Männchen vor der Paarung ihr Kopulationsorgan durch Vorkrümmen der Abdomenspitze mit der primären Geschlechtsöffnung am 9. Segment in Kontakt bringen und mit Sperma auffüllen. Dieses Auffüllen erfolgt entweder vor, sehr viel häufiger aber nach Ergreifen des Weibchens in Tandemposition und dauert 0,7 bis 76 Sekunden (CORBET 1999).

Nachdem das Erkämpfen eines Territoriums und die Partnerwahl am Rendezvousplatz abgelaufen sind, verankert das Männchen seine Hinterleibsanhänge am Kopf (Anisoptera) oder am Prothorax des Weibchens (Zygoptera) und bildet somit die Paarungskette oder das Tandem (Praecopula). Während in selteneren Fällen das Auffüllen des akzessorischen Kopulationsorgans bereits vor der Tandembildung erfolgt ist, geschieht dies vielfach nach Tandembildung mit angekoppeltem Weibchen. Bei der eigentlichen Copula krümmt das Weibchen sein Abdomenende zum akzessorischen Kopulationsorgan des Männchens und es erfolgt die Spermaübertragung, gekoppelt mit Spermakonkurrenz, wobei das einzigartige „Paarungsrad" der Odonaten gebildet wird. Dieses wird bei vielen Anisopteren direkt im Flug aus der Tandemstellung heraus gebildet, bei Zygopteren und Aeshnidae setzt sich das Tandem aber auch auf einer festen Unterlage ab und bildet dort das Paarungsrad. Anschließend trennen sich die Partner oder bleiben bis zur gemeinsamen Eiablage in Postcopula verbunden, wobei wieder die Tandemstellung eingenommen wird.

Die Eiablage des Weibchens erfolgt entweder allein, mit bewachendem, aber getrennt fliegendem Männchen oder in Postcopula mit angekoppeltem Männchen, wobei die Eiablagesequenz immer wieder durch nochmalige Kopulationen unterbrochen werden kann. Bei der Eiablage stechen Arten mit vollständig erhaltenem orthopteroiden Legeapparat (Zygoptera, Aeshnidae) ihre langgestreckten, bananenförmigen Eier einzeln, zu zweien (z. B. *Lestes sponsa*) oder zu vieren (z. B. *Lestes viridis*) in die Rinde von Ufergehölzen, Hochstauden, in emerse und submerse Wasserpflanzen oder, seltener, in (trockenen) Schlamm und Detritus, wobei manche Arten eine hohe Substratspezifität zeigen. Die Weibchen der Cordulegasteridae besitzen eine lange Legeröhre, mit der die Eier in Feinsand oder Schlammsubstrat von Bächen eingebracht werden (MÜLLER 1999). Alle übrigen einheimischen Arten haben keinen Ovipositor und legen ihre rundlichen Eier entweder in Gallertstränge eingebettet an Wasserpflanzen (*Epitheca bimaculata*), zu Klumpen vereinigt oder einzeln ab. Die Eier werden dabei direkt im Flug abgeworfen oder aber in einem anhaftenden Wassertropfen an der Abdomenspitze gesammelt und durch Eintauchen des Hinterleibs in das Wasser verbracht.

Die Eireife und –ablage bei Odonaten erfolgt in Schüben, die durch jeweils einige Tage voneinander getrennt sind. Geht die Eiablage ungestört vor sich, legen Weibchen den gesamten reifen Eivorrat (200–400 Eier bei *Coenagrion puella*, >2000 bei *Sympetrum fonscolombei*) auf einmal ab (CORBET 1999). Im Laufe ihres Lebens produziert *Coenagrion puella* 4–15 solcher Eischübe; die Zahl der produzierten Eier kann somit bei > 4000 liegen (BANKS & THOMPSON 1987, THOMPSON 1989) und beträgt bei *Pyrrhosoma nymphula* sogar ca. 8500 (BENNETT & MILL 1995).

Die einheimischen Großlibellenlarven können nach CORBET (1999) in schlanke, substrat- oder makrophytengebundene Formen mit ausgeprägter Thigmotaxis („claspers", z. B. Aeshnidae), in gedrungene, vagile, kriechende („sprawlers", z. B. Corduliidae, Libellulidae) und in stark beborstete, grabende, gedrungene Typen („shallow burrowers", z. B. Cordulegasteridae, Gomphidae) eingeteilt werden. Bei Kleinlibellenlarven sind die schlanken Calopterygidae mit ihren dreikantig-lamellaten Schwanzanhängen von den übrigen heimischen Arten mit vertikal-lamellaten Schwanzanhängen zu unterscheiden.

Die unter den Insekten einzigartige larvale Fangmaske stellt eine wesentliche Autapomorphie der Odonata dar. Dieses Organ ist eine Modifikation des Labiums und bietet den Libellenlarven einen wesentlich erweiterten Aktionsradius beim Beutefang. Im Ruhezustand ist die Fangmaske an die Kopfunterseite und zwischen den Coxen der Beine angelegt. Je nach Typ bedeckt dann die schöpflöffelartig gewölbte Fangmaske den ventralen Kopfbereich völlig (Typ der Helmmaske bei Cordulegasteridae, Corduliidae und Libellulidae) oder es bleiben die Kopfseiten, die übrigen Mundwerkzeuge und das Labrum frei sichtbar (Flachmaske bei allen übrigen einheimischen Familien; WARINGER 1981). Die Libellenlarven sind entweder ortsgebundene Lauerer oder vagile Jäger, die oft in der Nacht aktiv nach Beute suchen (CORBET 1999).

Die meisten Libellenlarven erfassen vor allem optisch ihre Beute. Viele Beobachtungen deuten zudem darauf hin, dass für einen erfolgreichen Fang ein Berührungskontakt der Antennen mit der Beute ebenso wichtig ist (WARINGER 1982). Befindet sich das Beutetier innerhalb der Reichweite der Fangmaske, wird diese durch Turgordruckerhöhung gestreckt vorgeschnellt und die Seitenlappen mit den Endhaken laterad abgespreizt. Helmmasken eignen sich vor allem für den Fang kleinerer Beuteorganismen, Flachmasken stellen hingegen eine Anpassung an grössere Beutetiere dar (PRITCHARD 1965). Allgemein ist das Beutespektrum der einheimischen Odonatenlarven sehr groß und deckt sich vielfach mit dem aktuellen Angebot des Wohngewässers. Die wichtigsten Beutetiere für Zygoptera stellen bei *Coenagrion puella* die Crustacea (Cladocera, Copepoda, Ostracoda), frühe Stadien von Wasserinsekten (z. B. Ephemeroptera, Odonata, Chironomidae), aber auch Wassermilben und frisch geschlüpfte oder juvenile Oligochaeta. Zusätzlich be-

33

wältigen Anisopterenlarven auch ältere Stadien von Oligochaeten und aquatischen Insektenlarven (Ephemeroptera, Odonata, Chironomidae, Simuliidae, Trichoptera, Megaloptera, Coleoptera), ferner Hirudinea, Mollusken, Krebse (Gammaridae, Isopoda) und sogar Fischbrut (SUHLING & MÜLLER 1996). Auch Wasserläufer (*Gerris*- und *Hydrometra*-Arten) sowie Kaulquappen stellen zu bestimmten Zeiten eine wichtige Futterquelle für Anisopterenlarven dar, wobei vor allem langsam schwimmende Krötenlarven wie z. B. die von *Bufo bufo* gern angenommen werden (CHOVANEC 1992a, b). Darüber hinaus ist wie bei den Imagines Kannibalismus unter Libellenlarven häufig.

Die Entwicklung der Larven verläuft bei Libellen (ohne Prolarvenstadium) über 7 bis 14 Stadien (ROBERT 1959, SUHLING & MÜLLER 1996). Das Larvenwachstum der Odonaten ist wie bei allen Insekten stark temperaturabhängig. Die Endlängen der einheimischen Libellenlarven schwanken zwischen 10,5 mm (inklusive Caudallamellen) bei *Nehalennia speciosa* bis zu 54 mm bei *Anax imperator* sowie 50 mm bei weiblichen Exemplaren von *Cordulegaster heros*, der massigsten und eindrucksvollsten Larve der österreichischen Odonatenfauna (HEIDEMANN & SEIDENBUSCH 1993; LANG et al. 2001). In unseren Breiten sind alle Kleinlibellenarten sowie *Sympetrum* spp. univoltin, *Libellula* spp., *Orthetrum* spp. und *Leucorrhinia* spp. semivoltin. *Aeshna cyanea*, *A. viridis*, *Gomphus pulchellus*, *G. vulgatissimus* und *Somatochlora* spp. benötigen 2–3 Jahre, *Gomphus flavipes*, *G. simillimus* und die beiden einheimischen *Onychogomphus*-Arten 3 Jahre, *Ophiogomphus cecilia* 3–4 Jahre, *Aeshna juncea* 4 Jahre und die *Cordulegaster*-Arten bis zu 5 Jahre (!) für eine Generation (LANG et al. 2001, PREWEIN 1996 sowie SUHLING & MÜLLER 1996).

3. Libellen als Bioindikatoren

Der Einsatz von Libellen als Bioindikatoren hat national und international eine lange Tradition und beruht auf der Korrelation zwischen dem Vorkommen bestimmter Arten(gesellschaften) und hydrologisch-morphologischen Habitatparametern sowie Vegetationsstrukturen (z. B. SCHMIDT 1985, 1989, WARINGER 1989, CHOVANEC & WARINGER 2001, SAHLEN & EKESTUBBE 2001, SCHINDLER et al. 2003, OERTLI 2008, SILVA et al. 2010). Aufgrund der Besiedlung verschiedener Teillebensräume sind Libellen ausgezeichnete Zeiger für den morphologischen Zustand der Gewässer und ihrer Uferbereiche sowie der Wasser-Land-Vernetzung (siehe dazu u. a. CHOVANEC & WARINGER 2007, SIMAIKA & SAMWAYS 2008, 2009, KUTCHER & BRIED 2014).

Libellen erfüllen alle essenziellen Anforderungen an Bioindikatoren: die ökologischen Ansprüche der heimischen Arten sind vergleichsweise sehr gut bekannt; Libellen reagieren unmittelbar auf Veränderungen ihrer Lebensräume und besiedeln daher aufgrund ihres Ausbreitungsverhaltens neue, geeignete Habitate; die Artenzahl ist überschaubar und die Imagines sind bereits im Feld am lebenden Tier bestimmbar (CHOVANEC & WARINGER 2007). Zugunsten von Libellen an Gewässern ergriffene Maßnahmen kommen großen Teilen der gesamten Lebensgemeinschaft zugute.

Libellen sind deshalb als „umbrella indicators" zu bezeichnen. Außerdem stellen Libellen aufgrund ihrer Auffälligkeit und Attraktivität die wohl „populärste" aquatische Insektengruppe dar („iconic, charismatic flagships"; vgl. dazu auch SAMWAYS 2008). Methoden und Ergebnisse von Renaturierungsmaßnahmen sowie Ziele von Schutzstrategien können deshalb der Öffentlichkeit anschaulich vermittelt werden.

In der österreichischen Wasserwirtschaft wurden bereits mehrfach Gewässerbewertungen auf der Grundlage libellenkundlicher Untersuchungen durchgeführt. Seit dem Inkrafttreten der EU Wasserrahmenrichtlinie standen hierbei die Entwicklung und Anwendung gewässertyp-bezogener Ansätze im Vordergrund. Beispielhaft seien hier folgende Studien angeführt: Bewertung neu geschaffener, stehender Gewässer und Uferstrukturen auf der Donauinsel in Wien (CHOVANEC & RAAB 2002, RAAB 2003), Evaluierung der Renaturierung an Mauerbach und Wienfluss in Wien (RAAB 2002), Bewertung des österreichischen Bodenseeufers in Vorarlberg (CHOVANEC et al. 2010), Erfolgskontrolle von Renaturierungsmassnahmen an mehreren Bächen im Weinviertel in Niederösterreich (CHOVANEC et al. 2012, 2014a, CHOVANEC 2014a) und an der Krems in Oberösterreich (CHOVANEC 2014b).

Als index- und gewässertyp-bezogene Methode zur Bewertung von Fluss-Au-Systemen wurde der Odonata-Habitat-Index (OHI; CHOVANEC & WARINGER 2001) entwickelt und an der gesamten österreichischen Donau eingesetzt (SCHULTZ et al. 2003, CHOVANEC et al. 2004). Auch in den o. g. Projekten wurde der OHI als eine der Bewertungsgrundlagen verwendet. Im Jahr 2014 wurde eine Methode zur libellenkundlichen Bewertung der Morphologie von Bächen und Flüssen der Bioregion Östliche Flach- und Hügelländer entwickelt, wobei der Bezug zu Planung und Evaluierung lokaler, an diesen Gewässern gesetzten Maßnahmen im Vordergrund stand. Große Flüsse (Donau, March, Thaya) waren nicht Gegenstand des Projektes. Die Fließgewässertypen dieser Bioregion sind in ihrer natürlichen Ausprägung – bedingt durch das überwiegend flache Gefälle – u. a. durch folgende Charakteristika geprägt: breite laterale Ausdehnung der Wasser-Land-Vernetzungszone, geringe Strömungsgeschwindigkeit, strömungsberuhigte Bereiche in den Hauptgerinnen, temporäre Vernässungsflächen, sumpfige Verlandungsbereiche. Libellen sind geeignete Indikatoren für die Gewässertypen dieser Bioregion, da sie die o. g. Teillebensräume besiedeln und damit eine Bewertung unter Einbeziehung sämtlicher lotischer und lenitischer, perennierender und temporärer Kompartimente erlauben (CHOVANEC et al. 2014b, 2015).

III Spezieller Teil

1. Systematik und Nomenklatur

Systematik und Nomenklatur folgen DIJKSTRA & LEWINGTON (2006), DIJKSTRA & KALKMAN (2012), DIJKSTRA et al. (2013) und WILDERMUTH & MARTENS (2014). Diesen Autoren zufolge sind für den Großteil der europäischen Taxa keine nomenklatorischen Änderungen mehr zu erwarten. Ausnahmen sind die generischen Zuordnungen von *Libellula depressa* und *Libellula fulva* (möglicherweise zur nearktischen Gattung *Ladona* zu stellen?) sowie von *Gomphus flavipes*, *Aeshna affinis*, *Aeshna isosceles*, *Aeshna mixta* und *Anax ephippiger*.

2. Angaben zu den Arten

Wissenschaftliche Namen, Synonyme

Bei allen Arten werden jene wissenschaftlichen Namen, die im Libellen-Teil des ersten Catalogus Faunae Austriae (ST. QUENTIN 1959) und in der späteren österreichischen faunistischen Literatur Verwendung fanden, ergänzend genannt.

Verbreitung

Unter „Verbr." werden jene Bundesländer angeführt, aus denen Nachweise der jeweiligen Art vorliegen. Ost- und Nordtirol werden aus zoogeographischen Gründen getrennt dargestellt. Die Abkürzungen bedeuten:

B	=	Burgenland
K	=	Kärnten
N	=	Niederösterreich
nT	=	Nordtirol
O	=	Oberösterreich
Ö	=	in allen Bundesländern und sowohl in Ost- als auch in Nordtirol vorkommend.
oT	=	Osttirol
S	=	Salzburg
St	=	Steiermark
V	=	Vorarlberg
W	=	Wien

Die weiteren Symbole bedeuten:

Ein nachgestelltes „(+)" weist darauf hin, dass kein aktuelles Vorkommen aus dem jeweiligen Bundesland mehr bekannt und die Art regional möglicherweise ausgestorben ist.

Wird das Bundesland in runden Klammern geführt, ist kein autochthones Vorkommen bekannt (z. B. nur Nachweise migrierender Tiere).

Wird das Bundesland in eckigen Klammern geführt, kommt die Art im Bundesland vermutlich nicht vor, historische Nachweise erfolgten vermutlich oder sicher irrtümlich.

Darüber hinaus werden nur jene Nachweise kommentiert, die Änderungen gegenüber der Checkliste in RAAB et al. (2006) darstellen.

3. Kommentiertes Verzeichnis der Libellen Österreichs

Ordnung Odonata FABRICIUS, 1793
Unterordnung Zygoptera SELYS, 1854

Überfamilie Lestoidea CALVERT, 1901
Familie Lestidae CALVERT, 1901

Gattung *Lestes* LEACH, 1815

Lestes barbarus (FABRICIUS, 1798)
Südliche Binsenjungfer
Verbr.: Ö

Lestes dryas KIRBY, 1890
Glänzende Binsenjungfer
Verbr.: Ö

Lestes macrostigma (EVERSMANN, 1836)
Dunkle Binsenjungfer
Verbr.: B, N, St, W

Lestes sponsa (HANSEMANN, 1823)
Gemeine Binsenjungfer
Verbr.: Ö

Lestes virens (CHARPENTIER, 1825)
Kleine Binsenjungfer
Verbr.: Ö (nicht in oT)
Anm.: Die Bestände in Österreich sind der mitteleuropäischen Unterart *L. virens* ssp. *vestalis* (RAMBUR, 1842) zuzurechnen.

Gattung *Chalcolestes* Kennedy, 1920

Chalcolestes parvidens (Artobolewskii, 1929)
Östliche Weidenjungfer
Verbr.: B, K
Anm.: Erstnachweis für Österreich durch Olias (2005) aus dem Burgenland.
Erstnachweis für Kärnten durch Brunner et al. (2013). Früher z.T. als Unterart von
Chalcolestes viridis geführt.

Chalcoestes viridis (Vander Linden, 1825)
Gemeine Weidenjungfer
Verbr.: Ö
Anm.: Früher in die Gattung *Lestes* gestellt.

Gattung *Sympecma* Burmeister, 1839

Sympecma fusca (Vander Linden, 1820)
Gemeine Winterlibelle
Verbr.: Ö (nicht in oT)

Sympecma paedisca (Brauer, 1877)
Sibirische Winterlibelle
Verbr.: nT, S(+), V
Anm.: Nur im Rheindelta und im Unterinntal.

Überfamilie Calopterygoidea Sélys, 1850

Familie Calopterygidae Sélys, 1850

Gattung *Calopteryx* Leach, 1815

Calopteryx splendens (Harris, 1782)
Gebänderte Prachtlibelle
Verbr.: Ö
Anm.: Die österreichischen Populationen zählen zur mitteleuropäischen Unterart
C. splendens ssp. *splendens*. An Fließgewässern.

Calopteryx virgo (Linnaeus, 1758)
Blauflügel-Prachtlibelle
Verbr.: Ö
Anm.: Die österreichischen Populationen zählen zur mitteleuropäischen Unterart
C. virgo ssp. *virgo*. An Fließgewässern.

Überfamilie Coenagrionidea KIRBY, 1890

Familie Platycnemididae YAKOBSON & BIANCHI, 1905

Gattung *Platycnemis* BURMEISTER, 1839

Platycnemis pennipes (PALLAS, 1771)
Verbr.: Ö (nicht in oT)
Anm.: Die Bestände in Österreich sind der mitteleuropäischen Unterart *P. pennipes* ssp. *pennipes* zuzurechnen.

Familie Coenagrionidae KIRBY, 1890

Anm.: Früher Agrionidae LEACH, 1815; dieser Name wird aufgrund von historischen nomenklatorischen Problemen (vgl. z. B. CALVERT et al. 1949) nicht mehr verwendet.

Gattung *Coenagrion* KIRBY, 1840

Anm.: Früher *Agrion* LEACH, 1815 nec FABRICIUS, 1775; dieser Name wird aufgrund von historischen nomenklatorischen Problemen (vgl. z. B. CALVERT et al. 1949) nicht mehr verwendet.

Coenagrion hastulatum (CHARPENTIER, 1825)
Speer-Azurjungfer
Verbr.: K, N, nT, O, S, St, V, W

Coenagrion hylas (TRYBOM, 1889)
Bileks Azurjungfer
Verbr.: nT
Anm.: Die österreichischen Tiere zählen zur Unterart *C. hylas* ssp. *freyi* BILEK, 1954. Da die Vorkommen in Bayern erloschen sind, sind die (wenigen) Populationen im nordtiroler Inn- und Lechtal die letzten bekannten Vorkommen dieser Unterart, deren Nominatform in Sibirien weit verbreitet ist.

Coenagrion lunulatum (CHARPENTIER, 1840)
Mond-Azurjungfer
Verbr.: N(+), nT, O, S(+?)
Anm.: In Lödl (1976a und b) sub *Coenagrion vernale* (HAGEN, 1839). Österreichweit sind gegenwärtig nur zwei aktuelle Vorkommen in Tirol bekannt.

Coenagrion mercuriale (CHARPENTIER, 1840)
Helm-Azurjungfer
Verbr.: nT(+), V
Anm.: Westeuropäisch verbreitete Art, nur im Westen Österreichs.

Coenagrion ornatum (SÉLYS, 1850)
Vogel-Azurjungfer
Verbr.: B, K(+?), N, S, St, V(+),W

Coenagrion puella (LINNAEUS, 1758)
Hufeisen-Azurjungfer
Verbr.: Ö

Coenagrion pulchellum (Vander Linden, 1825)
> Fledermaus-Azurjungfer
> Verbr.: Ö (nicht in oT)

Coenagrion scitulum (Rambur, 1842)
> Gabel-Azurjungfer
> Verbr.: B, N, St, W

Gattung *Enallagma* Charpentier, 1840

Enallagma cyathigerum (Charpentier, 1840)
> Gemeine Becherjungfer, Becherazurjungfer
> Verbr.: Ö

Gattung *Erythromma* Charpentier, 1840

Erythromma lindenii (Sélys, 1840)
> Pokaljungfer, Pokalazurjungfer
> Verbr.: B, N, nT, O, S, V
> Anm.: Früher in die Gattung *Cercion* Navás, 1907 oder *Coenagrion* Kirby, 1840 gestellt. In Kies- und Schottergruben mit sub- und emerser Vegetation, sehr selten.

Erythromma najas (Hansemann, 1823)
> Großes Granatauge
> Verbr.: Ö (nicht in oT)
> Anm.: An Stillgewässern mit dichter Schwimmblattvegetation.

Erythromma viridulum (Charpentier, 1840)
> Kleines Granatauge
> Verbr.: Ö (nicht in oT)
> Anm.: An Stillgewässern mit dichter Schwimmblattvegetation.

Gattung *Ischnura* Charpentier, 1840

Ischnura elegans (Vander Linden, 1820)
> Große Pechlibelle
> Verbr.: Ö
> Anm.: Weit verbreitet ist die Unterart *I. elegans* ssp. *elegans*, im Osten und Süden Österreichs kommt nach Schmidt (1967) und Stark (1976) auch die Unterart *I. elegans* ssp. *pontica* Schmidt, 1938 vor.

Ischnura pumilio (Charpentier, 1825)
> Kleine Pechlibelle
> Verbr.: Ö

Gattung *Nehalennia* Sélys in Sélys & Hagen, 1850

Nehalennia speciosa (Charpentier, 1840)
> Zwerglibelle
> Verbr.: (B), K, nT, O, St
> Anm.: In Zwischenmooren, sehr anspruchsvoll und selten.

Gattung *Pyrrhosoma* CHARPENTIER, 1840

Pyrrhosoma nymphula (SULZER, 1776)
>   Frühe Adonislibelle
>   Verbr.: Ö

Unterordnung Anisoptera SÉLYS, 1854

Überfamilie Aeshnoidea LEACH, 1815

Familie Aeshnidae LEACH, 1815

Gattung *Aeshna* FABRICIUS, 1775

>   Anm.: Manchmal auch *Aeschna* auctt., diese Schreibweise ist allerdings nicht korrekt (ungerechtfertigte Emendation).

Aeshna juncea (LINNAEUS, 1758)
>   Torf-Mosaikjungfer
>   Verbr.: K, N, nT, O, oT, S, St, V, W(+)

Aeshna affinis VANDER LINDEN, 1820
>   Südliche Mosaikjungfer
>   Verbr.: B, K, N, O, S, St, V, W
>   Erstnachweis aus Vorarlberg durch FRIEBE (2014)
>   Anm.: Migrationsfreudige, wärmeliebende Art.

Aeshna caerulea (STRÖM, 1783)
>   Alpen-Mosaikjungfer
>   Verbr.: K, nT, O, oT, S, St, V
>   Anm.: Von der montanen bis in die alpine Stufe verbreitet.

Aeshna cyanea (MÜLLER, 1764)
>   Blaugrüne Mosaikjungfer
>   Verbr.: Ö

Aeshna grandis (LINNAEUS, 1758)
>   Braune Mosaikjungfer
>   Verbr.: Ö
>   Erstnachweis aus dem Burgenland durch HÖTTINGER (2008)

Aeshna isoceles (MÜLLER, 1767)
>   Keilfleck-Mosaikjungfer
>   Verbr.: Ö (nicht in oT)
>   Anm.: Früher in die Gattung *Anaciaeschna* SÉLYS, 1878 gestellt.

Aeshna mixta LATREILLE, 1805
>   Herbst-Mosaikjungfer
>   Verbr.: Ö (nicht in oT)

Werner E. Holzinger et al.

Aeshna subarctica WALKER, 1908
Hochmoor-Mosaikjungfer
Verbr.: nT, N, O, S, St, V, W
Anm.: Die österreichischen Populationen zählen zur europäischen Unterart *A. subarctica* ssp. *elisabethae* DJAKONOV, 1922 (Syn.: *Aeshna subarctica* ssp. *interlineata* ANDER, 1944)

Aeshna viridis EVERSMANN, 1836
Grüne Mosaikjungfer
Verbr.: N
Anm.: In RAAB et al. (2006) wird die Bodenständigkeit noch als unsicher beurteilt, inzwischen liegen Belege für ein autochthones Vorkommen in den Donauauen oberhalb von Wien vor (Claus Stundner, pers. comm.). Zur Eiablage an Bestände von Krebsschere (*Stratiotes aloides*) gebunden.

Gattung *Anax* LEACH, 1815

Anax ephippiger (BURMEISTER, 1839)
Schabrackenlibelle, Schabracken-Königslibelle
Verbr.: B, N, O, S, W
Anm.: Früher in die Gattung *Hemianax* SÉLYS, 1883 gestellt.
Anm.: Migrationsfreudige, wärmeliebende Art.

Anax imperator LEACH, 1815
Große Königslibelle
Verbr.: Ö

Anax parthenope (SÉLYS, 1839)
Kleine Königslibelle
Verbr.: Ö (nicht in oT)

Gattung *Brachytron* EVANS, 1845

Brachytron pratense (MÜLLER, 1764)
Kleine Mosaikjungfer, Früher Schilfjäger
Verbr.: Ö (nicht in oT)

Überfamilie Gomphoidea RAMBUR, 1842
Familie Gomphidae RAMBUR, 1842
Gattung *Onychogomphus* SÉLYS, 1854

Onychgomphus forcipatus (LINNAEUS, 1758)
Kleine Zangenlibelle
Verbr.: Ö (nicht in oT, in V nach HOSTETTLER 2001 ausgestorben)
Anm.: Die Bestände in Österreich zählen zur Unterart *O. forcipatus* ssp. *forcipatus*.

Gattung *Ophiogomphus* SÉLYS, 1854

Ophiogomphus cecilia (GEOFFROY in FOURCROY, 1785)
    Syn.: *Ophiogomphus serpentinus* (CHARPENTIER, 1825)
    Grüne Flussjungfer
    Verbr.: B, K, N, O, S, St, W

Gattung *Gomphus* LEACH, 1815

Gomphus flavipes (CHARPENTIER, 1825)
    Asiatische Keiljungfer
    Verbr.: B, N, W

Gomphus pulchellus SÉLYS, 1840
    Westliche Keiljungfer
    Verbr.: V

Gomphus vulgatissimus (LINNAEUS, 1758)
    Gemeine Keiljungfer
    Verbr.: Ö (nicht in oT)

Überfamilie Cordulegastroidea HAGEN, 1875

Familie Cordulegastridae HAGEN, 1875

Gattung *Cordulegaster* LEACH, 1815

Cordulegaster bidentata SÉLYS, 1843
    Gestreifte Quelljungfer
    Verbr.: Ö
    Anm.: Die Bestände in Österreich sind der Unterart *C. bidentata* ssp. *bidentata* zuzu-
    rechnen. In Quellfluren und kleinen (Wald-)Bächen, wird oft übersehen.

Cordulegaster boltonii (DONOVAN, 1807)
    Zweigestreifte Quelljungfer
    Verbr.:K, N, nT, O, oT, S, St, V
    Anm.: Die Bestände in Österreich zählen zur Nominatform *C. boltonii* ssp. *boltonii*. In
    Quellfluren und kleinen (Wald-)Bächen, wird oft übersehen.

Cordulegaster heros THEISCHINGER, 1979
    Große Quelljungfer
    Verbr.: B, K, N, St, W
    Anm.: Die Art wurde aus Österreich (Niederösterreich (locus typicus: „St. Andrä
    v.d.H."), Wien, Steiermark) beschrieben; österreichischen Tiere zählen zur
    Nominatform *C. heros* ssp. *heros*. Vor THEISCHINGER (1979) wurden Tiere aus Österreich
    als *C. boltoni* ssp. *charpentieri* (KOLENATI, 1846) und gelegentlich auch *C. picta* SÉLYS,
    1854 identifiziert. Im Süden und Osten Österreichs in Quellfluren und kleinen (Wald-)
    Bächen, wird oft übersehen.

Überfamilie Libelluloidea LEACH, 1815

Familie Corduliidae SÉLYS, 1850

Gattung *Cordulia* LEACH, 1815

Cordulia aenea (LINNAEUS, 1758)
Falkenlibelle, Gemeine Smaragdlibelle
Verbr.: Ö

Gattung *Somatochlora* SÉLYS, 1871

Somatochlora alpestris (SÉLYS, 1840)
Alpen-Smaragdlibelle
Verbr.: K, NnT, O, oT, S, St, V

Somatochlora arctica (ZETTERSTEDT, 1840)
Arktische Smaragdlibelle
Verbr.: K, N, nT, O, oT, S, St,
Anm.: In montanen bis alpine Quellfluren und Mooren.

Somatochlora flavomaculata (VANDER LINDEN, 1825)
Gefleckte Smaragdlibelle
Verbr.: Ö
Erstnachweis aus dem Burgenland durch HÖTTINGER (2011)

Somatochlora meridionalis NIELSEN, 1935
Balkan-Smaragdlibelle
Verbr.: B, K, N, St
Erstnachweis aus dem Burgenland durch HÖTTINGER (2008)

Somatochlora metallica (VANDER LINDEN, 1825)
Glänzende Smaragdlibelle
Verbr.: Ö

Gattung *Epitheca* BURMEISTER, 1839

Epitheca bimaculata (CHARPENTIER, 1825)
Zweifleck
Verbr.: B, K, N, nT(+), O(+), S, St, W
An Seen mit gut ausgebildeter Schwimmblattzone; gute Bestände in Österreich nur an der Donau ab Wien sowie in den Marchauen.

Familie Libellulidae LEACH, 1815

Gattung *Crocothemis* BRAUER, 1868

Crocothemis erythraea (BRULLÉ, 1832)
Feuerlibelle
Verbr.: Ö (nicht in oT)
Anm.: Mediterrane Art, die sich erst in den letzten Jahrzehnten in Österreich etablieren konnte.

Gattung *Orthetrum* NEWMAN, 1833

Orthetrum albistylum (SÉLYS, 1848)
Östlicher Blaupfeil
Verbr.: Ö (nicht in nT)

Orthetrum brunneum (FONSCOLOMBE, 1837)
Südlicher Blaupfeil
Verbr.: Ö

Orthetrum cancellatum (LINNAEUS, 1758)
Großer Blaupfeil
Verbr.: Ö (nicht in oT)

Orthetrum coerulescens (FABRICIUS, 1798)
Kleiner Blaupfeil
Verbr.: Ö

Gattung *Libellula* LINNAEUS, 1758

Libellula depressa LINNAEUS, 1758
Plattbauch
Verbr.: Ö

Libellula fulva MÜLLER, 1764
Spitzenfleck
Verbr.: Ö (nicht in oT)

Libellula quadrimaculata LINNAEUS, 1758
Vierfleck
Verbr.: Ö

Gattung *Leucorrhinia* BRITTINGER, 1850

Leucorrhinia albifrons (BURMEISTER, 1839)
Östliche Moosjungfer
Verbr.: K, O(+), oT(+)
Anm.: Durch Lebensraumverluste unmittelbar vom Aussterben bedroht, österreichweit nur mehr ein Vorkommen in Südkärnten (HOLZINGER & KOMPOSCH 2012).

Leucorrhinia caudalis (C<small>HARPENTIER</small>, 1840)

Zierliche Moosjungfer
Verbr.: K, O(+),W(+)
Anm: Anspruchsvolle Art von Auen-Stillgewässern, die in Österreich ebenfalls vom Aussterben bedroht ist. Die letzten Nachweise gelangen in den Jahren 2000 (M. & F. Stich im Rosental, Kärnten) und 2003 (R. Raab in Wien; R<small>AAB</small> 2006).

Leucorrhinia dubia (V<small>ANDER</small> L<small>INDEN</small>, 1825)

Kleine Moosjungfer
Verbr.: K, N, nT, O, oT, S, St, V

Leucorrhinia pectoralis (C<small>HARPENTIER</small>, 1825)

Große Moosjungfer
Verbr.: B, K, N, nT, O, S, St, W

Leucorrhinia rubicunda (L<small>INNAEUS</small>, 1758)

Nordische Moosjungfer
Verbr.: N, nT(+), [?O], [?S], V(+)
Anm.: Boreale Art, nur in Mooren des Waldviertels vorkommend.

Gattung *Sympetrum* N<small>EWMAN</small>, 1833

Sympetrum danae (S<small>ULZER</small>, 1776)

Schwarze Heidelibelle
Verbr.: Ö

Sympetrum depressiusculum (S<small>ÉLYS</small>, 1841)

Sumpf-Heidelibelle
Verbr.: Ö (nicht in oT)
Anm.: Hochgradig gefährdete Art von Überschwemmungsflächen (Auwiesen), gute Bestände nur mehr aus dem Rheintal bekannt.

Sympetrum flaveolum (L<small>INNAEUS</small>, 1758)

Gefleckte Heidelibelle
Verbr.: Ö

Sympetrum fonscolombii (S<small>ÉLYS</small>, 1840)

Frühe Heidelibelle
Verbr.: Ö

Sympetrum meridionale (S<small>ÉLYS</small>, 1841)

Südliche Heidelibelle
Verbr.: B, N, (nT), O, (oT), St, W
Migrierationsfreudige thermophile Art mit Verbreitungsschwerpunkt im pannonischen Osten Österreichs.

Sympetrum pedemontanum (M<small>ÜLLER</small> in A<small>LLIONI</small>, 1766)

Gebänderte Heidelibelle
Verbr.: Ö

Sympetrum sanguineum (M<small>ÜLLER</small>, 1764)

Blutrote Heidelibelle
Verbr.: Ö

Sympetrum striolatum (CHARPENTIER, 1840)
> Große Heidelibelle
> Verbr.: Ö

Sympetrum vulgatum (LINNAEUS, 1758)
> Gemeine Heidelibelle
> Verbr.: Ö

4. Irrtümliche und fragliche Meldungen

Für insgesamt sieben Libellenarten existieren irrtümliche Meldungen in der Literatur. Sie werden alle von RAAB et al. (2006: 257 ff.) ausführlich kommentiert, daher werden sie hier nur kurz zusammengefasst wiedergegeben:

Epallage fatime (CHARPENTIER, 1840), Fam. Euphaeidae: Von ASKEW (1988) irrtümlich für Österreich gemeldet.

Platycnemis latipes RAMBUR, 1842: Irrtümliche Angaben in älteren Quellen, die auf BRAUER (1876, 1878) zurückzuführen sind.

Ceriagrion tenellum (DE VILLERS, 1789): Historische Meldungen (vgl. RAAB et al. 2006: 257 f.) sind unbelegt, ein Vorkommen in Österreich wäre allerdings denkbar.

Pyrrhosoma elisabethae SCHMIDT, 1948: STARK (1976, 1979) meldet intermediäre Formen zwischen *P. nymphula* und *P. elisabethae* aus Südostösterreich. Die Verbreitung letzterer Art ist ungenügend bekannt, bisher liegen nur wenige Fundmeldungen aus Albanien und Griechenland vor. Dennoch ist ein Vorkommen von *P. elisabethae* in Österreich nicht sehr wahrscheinlich.

Gomphus simillimus SÉLYS, 1840: Irrtümlich bei JANECEK et al. (1995) für Österreich angeführt.

Onychogomphus uncatus (CHARPENTIER, 1840): Historische bis relativ rezente (JANECEK et al. 1995) Meldungen, allerdings stets ohne Beleg. Ein Vorkommen ist nicht wahrscheinlich.

Cordulegaster picta SÉLYS, 1854: In älterer Literatur mehrfach irrtümlich für Österreich gemeldet (vgl. auch *Cordulegaster heros*).

5. Exotische Arten in Gewächshäusern

LAISTER et al. (2014) dokumentieren Nachweise von 23 exotischen (vorwiegend südostasiatischen) Libellenarten, die von 1991 bis 2011 in insgesamt mehr als 1.000 adulten Individuen in einem Gewächshaus für Aquarienpflanzen bei Wels (Oberösterreich) gefangen werden konnten. Regelmäßig und in größerer Zahl erfolgten Funde von *Ischnura senegalensis* (RAMBUR, 1842), *Pseudagrion microcepha-*

lum (RAMBUR, 1842), *Crocothemis servilia* (DRURY, 1773), *Neurothemis fluctuans* (FABRICIUS, 1793) und *Orthetrum sabina* (DRURY, 1770).

IV Literatur

ASKEW, R. R. 2004: The dragonflies of Europe. — Colchester: Harley Books, 308 pp.

BANKS, M. J., THOMPSON, D. J. 1987: Lifetime reproductive success of females of the damselfly *Coenagrion puella*. — Journal of Animal Ecology **56**: 815–832.

BELLMANN, H. 1987: Libellen beobachten – bestimmen. — Melsungen: Neumann-Neudamm, 272 S.

BENNETT, S., MILL, P. J. 1995: Lifetime egg production and egg mortality in the damselfly *Pyrrhosoma nymphula* (SULZER) (Zygoptera: Coenagrionidae). — Hydrobiologia **310**: 71–78.

BRAUER, F. 1876: Die Neuropteren Europas und insbesondere Österreichs mit Rücksicht auf die geographische Verbreitung. — Festschrift zur Feier des 25-jährigen Bestehens der k. k. zoologisch-botanischen Gesellschaft Wien: 263–300.

BRAUER, F. 1878: Verzeichnis der Neuropteren Deutschlands und Österreichs. — Entomologische Nachrichten **4(6/7)**: 69–74 und 85–90.

BRUNNER, H., HOLZINGER, W. E., KOMPOSCH, B. 2013: Die Östliche Weidenjungfer (*Lestes parvidens*) neu für Kärnten, mit Ergänzungen und Korrekturen zu den „Libellen Kärntens". — Carinthia II **(203/123)**: 343–348.

CALVERT, P., LONGFIELD, C., COWLEY, J., SCHMIDT, E. 1949: *Agrion* versus *Calopteryx*. — Entomological News **60**: 145–151.

CHOVANEC, A. 1992a: Beutewahrnehmung (Reaktive Distanzen) und Beuteverfolgung (Kritische Distanzen) bei Larven von *Aeshna cyanea* (Müller) (Anisoptera: Aeshnidae). — Odonatologica **21**: 327–333.

CHOVANEC, A. 1992b: The influence of tadpole swimming behaviour on predation by dragonfly nymphs. — Amphibia-Reptilia **13**: 341–349.

CHOVANEC, A. 2009: Odonata (Libellen). — In RABITSCH, W., ESSL, F. (Hrsg.): Endemiten – Kostbarkeiten in Österreichs Pflanzen- und Tierwelt. — Naturwissenschaftlicher Verein für Kärnten und Umweltbundesmt GmbH, Klagenfurt und Wien, S. 592–593.

CHOVANEC, A. 2014a: *Coenagrion ornatum* (Selys, 1850) und *Ophiogomphus cecilia* (Fourcroy, 1785) (Insecta: Odonata) – Nachweis von zwei FFH-Arten an der Zaya (Niederösterreich). — Beiträge zur Entomofaunistik **14**: 1–11.

CHOVANEC, A. 2014b: Libellen als Indikatoren für den Erfolg von Renaturierungsmaßnahmen an Fließgewässern am Beispiel der Krems im Bereich Ansfelden/Oberaudorf. — ÖKO·L **36/2**: 17–26.

CHOVANEC, A., RAAB, R. 2002: Die Libellenfauna (Insecta: Odonata) des Tritonwassers auf der Donauinsel in Wien – Ergebnisse einer Langzeitstudie, Aspekte der Gewässerbewertung und Bioindikation. — Denisia **3**: 63–79.

CHOVANEC, A., WARINGER, J. 2001: Ecological integrity of river-floodplain systems – assessment by dragonfly surveys (Insecta: Odonata). — Regulated Rivers: Research, Management **17**: 493–507.

CHOVANEC, A., WARINGER, J. 2007: Libellen als Bioindikatoren. — In RAAB, R., CHOVANEC, A., PENNERSTORFER, J.: Libellen Österreichs. Springer: Wien, New York, S. 311–324.

CHOVANEC, A., SCHINDLER, M., PALL, K., HOSTETTLER, K. 2010: Bewertung des österreichischen Bodenseeufers auf der Grundlage libellenkundlicher Untersuchungen. — Schriftenreihe Lebensraum Vorarlberg, Bd. **59**.

CHOVANEC, A., SCHINDLER, M., RUBEY, W. 2014a: Assessing the success of lowland river restoration using dragonfly assemblages (Insecta: Odonata). — Acta ZooBot Austria **150/151**: 1–16.

CHOVANEC, A., SCHINDLER, M., WARINGER, J., WIMMER, R. 2015: The Dragonfly Association Index (Insecta: Odonata) – a tool for the type-specific assessment of lowland rivers. — River Research and Applications: DOI: 10.1002/rra.2760.

CHOVANEC, A., WARINGER, J., RAAB, R., LAISTER, G. 2004: Lateral connectivity of a fragmented large river system: assessment on a macroscale by dragonfly surveys (Insecta: Odonata). Aquatic Conservation: — Marine and Freshwater Ecosystems **14**: 163–178.

CHOVANEC, A., WARINGER, J., WIMMER, R., SCHINDLER, M. 2014b: Dragonfly Association Index – Bewertung der Morphologie von Fliessgewässern der Bioregion Östliche Flach- und Hügelländer durch libellenkundliche Untersuchungen. — Bundesministerium für Land- und Forstwirtschaft, Umwelt und Wasserwirtschaft.

CHOVANEC, A., WIMMER, R., RUBEY, W., SCHINDLER, M., WARINGER, J. 2012: Hydromorphologische Leitbilder als Grundlage für die Ableitung gewässertyp-spezifischer Libellengemeinschaften (Insecta: Odonata), dargestellt am Beispiel der Bewertung der restrukturierten Weidenbach-Mündungsstrecke (Marchfeld, Niederösterreich). — Wissenschaftliche Mitteilungen aus dem Niederösterreichischen Landesmuseum **23**: 83–112.

CORBET, P. S. 1999: Dragonflies. Behaviour and Ecology of Odonata. — Colchester: Harley Books, 829 pp.

CORBET, P. S., HARVEY, I. F., ABISGOLD, J., MORRIS, F. 1989: Seasonal regulation in *Pyrrhosoma nymphula* (Sulzer) (Zygoptera: Coenagrionidae). 2. Effect of photoperiod on larval development in spring and summer. — Odonatologica **18**: 333–348.

DIJKSTRA, K.-D. B., KALKMAN, V. J. 2012: Phylogeny, classification and taxonomy of European dragonflies and damselflies (Odonata: a review. — Organisms Diversity and Evolution **12** (3): 209–227.

DIJKSTRA, K.-D. B., LEWINGTON, R. 2006: Field Guide to the Dragonflies of Britain and Europe. — British Wildlife Publishing, 320 S.

DIJKSTRA, K-D. B., BECHLY, G., BYBEE, S. M., DOW, R. A., DUMONT, H. J., FLECK, G., GARRISON, R. W., HÄMÄLÄINEN, M., KALKMAN, V. J., KARUBE, H., MAY, M. L., ORR, A. G., PAULSON, D. R., REHN, A. C., THEISCHINGER, G., TRUEMAN, J. W. H., VAN TOL, J., VON ELLENRIEDER, N., WARE, J. 2013: The classification and diversity of dragonflies and damselflies (Odonata). — Zootaxa **3703** (1): 36–45.

DREYER, W. 1986: Die Libellen. — Hildesheim: Gerstenberg, 219 S.

Werner E. Holzinger et al.

ERIKSEN, C. H. 1986: Respiratory roles of caudal lamellae (gills) in a lestid damselfly (Odonata: Zygoptera). — Journal of the North American Benthological Society **5**: 16–27.

FRIEBE, G. 2014: Libellen-Beobachtungen (Einzelfunde) aus Vorarlberg (Odonata / Österreich – Austria occ.). — inatura Forschung online **9**: 1–13.

GORB, S. N. 1994: Female perching behaviour in *Sympetrum sanguineum* (MÜLLER) at feeding places (Anisoptera: Libellulidae). — Odonatologica **23**: 341–353.

HEIDEMANN, H., SEIDENBUSCH, R. 1993: Die Libellenlarven Deutschlands und Frankreichs. Handbuch für Exuviensammler. — Keltern: Verlag Erna Bauer, 391 S.

HOLZINGER W. E., KOMPOSCH B. (2012): Die Libellen Kärntens. — Sonderreihe Natur Kärnten, Bd. 6. Naturwissenschaftlicher Verein für Kärnten, Klagenfurt, 336 S.

HOSTETTLER, K. 2001: Libellen (Odonata) in Vorarlberg (Österreich). — Vorarlberger Rundschau **9**: 9–134.

HÖTTINGER, H. 2008: Nachweise der Baunen Mosaikjungfer *Aeshna grandis* (LINNAEUS, 1758) und der Balkan-Smaragdlibelle *Somatochlora meridionalis* (NIELSEN, 1935) aus dem Burgenland, östliches Österreich (Insecta: Odonata). — Beiträge zur Entomofaunistik **9**: 181–186.

HÖTTINGER, H. 2011: Erstnachweis der Gefleckten Smaragdlibelle *Somatochlora flavomaculata* (VANDER LINDEN, 1825) aus dem Burgenland, östliches Österreich (Insecta: Odonata). — Beiträge zur Entomofaunistik **12**: 123–127.

JANECEK, B., MOOG, O., WARINGER, J. 1995: Libellen (Odonata). — In MOOG, O. (Hrsg.): Fauna Aquatica Austriaca, Lieferung Mai/95. — Wasserwirtschaftskataster, Bundesministerium für Land- und Forstwirtschaft, Wien, 13 S.

JÖDICKE, R. 1997: Die Binsenjungfern und Winterlibellen Europas. — Die Neue Brehm-Bücherei, Bd. 631. Westarp Wissenschaften, Magdeburg, 277 S.

KAISER, H. 1974: Verhaltensgefüge und Temporalverhalten der Libelle *Aeshna cyanea* (Odonata). — Zeitschrift für Tierpsychologie **34**: 398–429.

KAISER, H. 1985: Availability of receptive females at the mating place and mating chances of males in the dragonfly *Aeschna cyanea*. — Behavioral Ecology and Sociobiology **18**: 1–7.

KIPPING, J. 2006: Globalisierung und Libellen: Verschleppung von exotischen Libellenarten nach Deutschland (Odonata: Coenagrionidae, Libellulidae). — Libellula **25** (1/2): 109–116.

KUTCHER, T.E., BRIED, J.T. 2014: Adult Odonata conservatism as an indicator of freshwater wetland condition. — Ecological Indicators **38**: 31–39.

LAISTER, G. 1996a: Bestand, Gefährdung und Ökologie der Libellenfauna der Großstadt Linz. — Naturkundliches Jahrbuch der Stadt Linz **40/41**, 1994/95: 9–305.

LAISTER, G. 1996b: Verbreitungsübersicht und eine vorläufige Rote Liste der Libellen Oberösterreichs. — Naturkundliches Jahrbuch der Stadt Linz **40/41**, 1994/95: 307–388.

LAISTER, G., LEHMANN, G., MARTENS, A. 2014: Exotic Odonata in Europe. Odonatologica **43(1/2)**: 125–135.

LANDMANN, A., LEHMANN, G., MUNGENAST, F., SONNTAG, H. 2005: Die Libellen Tirols. — Berenkamp, Wattens, 324 S.

LANG, C., MÜLLER, H., WARINGER, J. A. 2001: Larval habitats and longitudinal distribution patterns of *Cordulegaster heros* THEISCHINGER and *C. bidentata* SELYS in an Austrian forest stream (Anisoptera: Cordulegastridae). — Odonatologica **30**: 395–409.

LAWTON, J. H. 1971: Ecological energetics studies on larvae of the damselfly *Pyrrhosoma nymphula* (SULZER)(Odonata: Zygoptera). — Journal of Animal Ecology **40**: 385–423.

LEHMANN, G. 1994: Biometrische Veränderungen der Imagines von *Platycnemis pennipes* (PALLAS, 1771) im Verlauf einer Saison (Odonata: Platycnemididae). — Unveröff. Diplomarbeit an der Technischen Universität Braunschweig, 87 S.

LÖDL, M. 1976a: Die Libellenfauna Österreichs. — Linzer biologische Beiträge **8** (1): 13–159.

LÖDL, M. 1976b: Die Libellenfauna Österreichs, 1. Nachtrag. — Linzer biologische Beiträge **8** (2): 383–387.

MARTENS, A. 1996: Die Federlibellen Europas. — Die Neue Brehm-Bücherei, Bd. 626. Westarp Wissenschaften, Magdeburg, 149 S.

MARTENS, A., WIMMER, W. 1996: Schwärmende Ameisen (Hymenoptera: Formicidae) als Beute von Grosslibellen (Anisoptera: Aeshnidae). — Libellula **15**: 197–202.

MOORE, N. W. 1991: Male *Sympetrum striolatum* (CHARP.) „defends" a basking spot rather than a particular locality (Anisoptera: Libellulidae). — Notulae Odonatologicae **3**: 112.

MÜLLER, H. 1999: Schlüpfortwahl, Phänologie und Verteilungsmuster der Imagines von *Cordulegaster bidentata* SELYS, 1843 und *Cordulegaster heros* THEISCHINGER, 1979 im Einzugsgebiet des Weidlingbaches (Niederösterreich). — Unveröff. Diplomarbeit an der Universität Wien, 89 S.

OERTLI, B. 2008: The use of dragonflies in the assessment and monitoring of aquatic habitats. — In CÓRDOBA-AGUILAR, A. (Ed.): Dragonflies and Damselflies. Model Organisms for Ecological and Evolutionary Research. — New York: Oxford University Press, p. 79–95.

OLIAS, M. 2005: *Lestes parvidens* am Südostrand Mitteleuropas: Erste Nachweise aus Österreich, der Slowakei, Ungarn und Rumänien (Odonata: Lestidae). — Libellula **24** (3/4): 155–161.

PARR, M. J. 1970: The life histories of *Ischnura elegans* (VAN DER LINDEN) and *Coenagrion puella* (L.) (Odonata) in south Lancashire. — Proceedings of the Royal Entomological Society of London **45**: 172–181.

PARR, M. J. 1983: Some aspects of territoriality in *Orthetrum coerulescens* (FABRICIUS) (Anisoptera: Libellulidae). — Odonatologica **12**: 239–257.

PODA, N. 1761: Insecta musei graecensis. Graecium, 133+12 S.

PREWEIN, B. 1996: Ökologische Untersuchungen an Moorlibellen (Insecta: Odonata) im Rotmösl (Niederösterreich). — Unveröff. Diplomarbeit an der Universität Wien, 112 S.

Werner E. Holzinger et al.

PRITCHARD, G. 1965: Prey capture by dragonfly larvae. Canadian Journal of Zoology **43**: 271–289.

RAAB, R. 1994: Bibliographie zur Libellenfauna Österreichs. — Anax **1** (1): 10–23.

RAAB, R. 2003: Die Besiedlung neu geschaffener Uferstrukturen im Stauraum Freudenau (Wien, Niederösterreich) durch Libellen (Insecta, Odonata). — Denisia **10**: 79–99.

RAAB, R., CHOVANEC, A., PENNERSTORFER, J. 2007: Atlas der Libellen Österreichs. — Umweltbundesamt Wien. — Wien: Springer, 343 S.

RAAB, R., CHWALA, E. 1997: Libellen (Insecta: Odonata). Eine Rote Liste der in Niederösterreich gefährdeten Arten. — Amt der Niederösterreichischen Landes-regierung, Abteilung Naturschutz, Wien, 91 S.

RAAB, R. 2002: Libellen als Bioindikatoren zur Überprüfung der Effizienz von Revitalisierungsmassnahmen an Wienfluss und Mauerbach. — Perspektiven **1/2**: 55–62.

REHFELDT, G. E., KESERÜ, E., WEINHEBER, N. 1993: Opportunistic exploitation of prey in the libellulid dragonfly *Orthetrum cancellatum* (Odonata: Libellulidae). — Zoologische Jahrbücher (Systematik) **120**: 441–451.

ROBERT, P.-A. 1959: Die Libellen (Odonaten). — Naturkundliche Taschenbücher, Bd.4. — Bern: Kümmerly und Frey, 404 S.

SAHLEN, G., EKESTUBBE, K. 2001: Identification of dragonflies (Odonata) as indicators of general species richness in boreal forest lakes. — Biodiversity and Conservation **10**: 673–690.

SAMWAYS, M.J. 2008: Dragonflies as focal organisms in contemporary conservation bi-ology. — In CÓRDOBA-AGUILAR, A. (Ed.): Dragonflies and Damselflies. Model Organisms for Ecological and Evolutionary Research. — New York: Oxford University Press, p. 97–108.

SCHIEMENZ, H. 1953: Die Libellen unserer Heimat. — Jena: Urania-Verlag, 154 S.

SCHINDLER, M., FESL, C., CHOVANEC, A. 2003: Dragonfly associations (Insecta: Odonata) in relation to habitat variables: a multivariate approach. — Hydrobiologia **497**: 169–180.

SCHMIDT, E. 1967: Versuch einer Analyse der *Ischnura elegans*-Gruppe (Odonata, Zygoptera). — Entomologisk Tidskrift **88**: 188–225.

SCHMIDT, E. 1985: Habitat inventarization, characterization and bioindication by a „rep-resentative spectrum of Odonata species (RSO)". — Odonatologica **14**: 127–133.

SCHMIDT, E. 1989: Libellen als Bioindikatoren für den praktischen Naturschutz: Prinzipien der Geländearbeit und ökologischen Analyse und ihre theoretische Grundlegung im Konzept der ökologischen Nische. — Schr.-R. f. Landschaftspflege und Naturschutz **29**: 281–289.

SCHULTZ, H., WARINGER, J., CHOVANEC, A. 2003: Assessment of the ecological status of Danubian floodplains at Tulln (Lower Austria) based on the Odonata Habitat Index (OHI). — Odonatologica **32** (4): 355–370.

SILVA, D.P., DE MARCO, P., RESENDE, D.C. 2010: Adult odonate abundance and community assemblage measures as indicators of stream ecological integrity: a case study. — Ecological Indicators **10**: 744–752.

SIMAIKA J.P., SAMWAYS M.J. 2008: Valuing dragonflies as service providers. In CÓRDOBA-AGUILAR, A. (Ed.): Dragonflies and Damselflies. — Model Organisms for Ecological and Evolutionary Research. — New York: Oxford Univ. Press, p. 109–123.

SIMAIKA, J.P., SAMWAYS, M.J. 2009: An easy-to-use index of ecological integrity for prioritizing streams for conservation action. — Biodiversity and Conservation **18**: 1171–1185.

SONNTAG, H. 1999: Schlupfbiologische Freilanduntersuchungen an Libellen unter besonderer Berücksichtigung von Sympecma paedisca (BRAUER, 1877) (Insecta: Odonata). — Diplomarbeit Universität Innsbruck, 119 S.

ST. QUENTIN, D. 1959: Odonata. — Catalagus Faunae Austriae **12c**: 1–11.

STARK, W. 1979: Mischformen von Pyrrhosoma n. nymphula (SULZER, 1776) und Pyrrhosoma n. elisabethae SCHMIDT, 1948 aus der Steiermark, Österreich (Zygoptera: Coenagrionidae). — Notulae Odonatologicae **1/4**: 53–54.

STARK, W. 1976: Die Libellen der Steiermark und des Neusiedlerseegebietes in monographischer Sicht. — Dissertation an der Karl-Franzens-Universität Graz, 186 S.

STERNBERG, K. 1990: Autökologie von sechs Libellenarten der Moore und Hochmoore des Schwarzwaldes und Ursachen ihrer Moorbindung. — Dissertation an der Universität Freiburg, 431 S.

STERNBERG, K. 1995: Influence of oviposition date and temperature upon embryonic development in Somatochlora alpestris and S. arctica (Odonata: Corduliidae). — Journal of Zoology (London) **235** (1): 163–174.

SUHLING, F., MÜLLER, O. 1996: Die Flussjungfern Europas. — Die Neue Brehm-Bücherei, Bd. **628**. — Magdeburg: Westarp Wissenschaften, 237 S.

SUKHACHEVA, G. A. 1996: Study of the natural diet of adult dragonflies using an immunological method. — Odonatologica **25**: 397–403.

THEISCHINGER, G. 1979: Cordulegaster heros sp. nov. und Cordulegaster heros pelionensis ssp. nov., zwei neue Taxa des Cordulegaster boltoni (DONOVAN)-Komplexes aus Europa (Anisoptera: Cordulegasteridae). — Odonatologica **8**: 23–38.

THOMPSON, D. J. 1989: A population study of the azure damselfly Coenagrion puella (L.) in northern England. — Journal of the British Dragonfly Society **5**: 17–22.

UTZERI, C., FALCHETTI, E., CARCHINI, G. 1976: Alcuni aspetti etologici della ovideposizione di Lestes barbarus (FABRICIUS) presso pozze temporanee (Zygoptera: Lestidae). — Odonatologica **5**: 175–179.

WAAGE, J. K. 1979: Dual function of the damselfly penis: sperm removal and transfer. — Science **203**: 916–917.

WARINGER, J. 1981: Die Fangmaske der Libellenlarven: Anisoptera (Grosslibellen). — Mikrokosmos **70**: 266–270.

WARINGER, J. 1982: Die Fangmaske der Libellenlarven: Kleinlibellen. — Mikrokosmos **71**: 118–122.

Werner E. Holzinger et al.

WARINGER, J. 1989: Gewässertypisierung anhand der Libellenfauna am Beispiel der Althenwörther Donauau (Niederösterreich). — Natur und Landschaft **64**: 389–392.

WARINGER, J. 2007: Biologie der Libellen. — In RAAB, R., CHOVANEC, A., PENNERSTORFER, J. (eds.): Libellen Österreichs. — Wien, New York: Springer, S. 5–34.

WARINGER, J. A., HUMPESCH, U. H. 1984: Embryonic development, larval growth and life cycle of *Coenagrion puella* (Odonata: Zygoptera) from an Austrian pond. — Freshwater Biology **14**: 385–399.

WENDLER, A. 1994: Ökologische Betrachtungen zur Dauer der Eientwicklung von *Platycnemis pennipes* (PALLAS, 1771) unter Laborbedingungen mir Hilfe mathematischer Modelle (Odonata: Platycnemididae). — Unveröff. Diplomarbeit an der Universität Braunschweig, 89 S.

WENDLER, A., NÜSS, J.-H. 1991: Libellen. — Hamburg: DJN, 129 S.

WILDERMUTH, H. 1994: Populationsdynamik der Großen Moosjungfer, *Leucorrhinia pectoralis* Charpentier, 1825 (Odonata: Libellulidae). — Zeitschrift für Ökologie und Naturschutz **3**: 25–39.

WILDERMUTH, H., MARTENS, A. 2014: Taschenlexikon der Libellen Europas. — Wiebelsheim: Quelle & Meyer, 824 S.

Anschriften der Verfasser:

Priv.-Doz. Dr. Werner E. HOLZINGER
Ökoteam – Institut für Tierökologie und Naturraumplanung
Bergmanngasse 22, A-8010 Graz, Austria
E-Mail: holzinger@oekoteam.at

Univ.-Prof. Dr. Johann A. WARINGER
Department für Limnologie und Ozeanographie
Universität Wien, Althanstrasse 14, A-1090 Wien, Austria
E-Mail: johann.waringer@univie.ac.at

Univ.-Doz. Dr. Andreas CHOVANEC
Umweltbundesamt Wien
Spittelauer Lände 5, A-1090 Wien, Austria
E-Mail: andreas.chovanec@umweltbundesamt.at